Collins **AQA GCS** **n**

G000273937

Grade Booster

Combined Science

Trilogy

Tom Adams, Dan Evans and Dan Foulder

Contents

Contents

Acknowledgements

The authors and publisher are grateful to the copyright holders for permission to use quoted materials and images.

Every effort has been made to trace copyright holders and obtain their permission for the use of copyright material. The authors and publisher will gladly receive information enabling them to rectify any error or omission in subsequent editions. All facts are correct at time of going to press.

Cover and p1 © Shuttertstock/SpeedKingz, © Shuttertstock/ Marina Sun, © Shutterstock/Omelchenko

All other images are © Shutterstock.com and ©HarperCollins*Publishers*

Published by Collins
An imprint of HarperCollins*Publishers*
1 London Bridge Street
London SE1 9GF

ISBN: 978-0-00-827684-3

First published 2018
10 9 8 7 6 5 4 3 2 1
© HarperCollins*Publishers* Limited 2018

British Library Cataloguing in Publication Data.

A CIP record of this book is available from the British Library.

Commissioning Editor: Kerry Ferguson
Authors: Tom Adams, Dan Evans and Dan Foulder
Project Editor: Charlotte Christensen
Project Manager and Editorial: Jill Laidlaw
Cover Design: Sarah Duxbury
Inside Concept Design: Paul Oates
Text Design and Layout: QBS Learning
Production: Natalia Rebow
Printed in Great Britain by Martins the Printers

Introduction

About this Book

This book has been designed to support your preparation for the AQA GCSE Combined Science: Trilogy (9–1) examination and help you achieve your best possible grade.

The AQA Combined Science: Trilogy specification is divided into 24 topics, seven biology, 10 chemistry and seven physics, although not all of the topics are equal in size. This Grade Booster book mirrors that approach and the worked questions are divided into the same topics as on the specification, so you can easily find questions that cover every topic you will study.

As you revise each topic in this book, you will find exam-style questions, model answers and supporting notes with tips and hints. You will also find guidance on what the examiner is looking for and revision advice for different parts of the specification. Questions that only cover Higher Tier content are shown by this symbol: **HT**.

You can visit the AQA website to download or view a copy of the GCSE Combined Science: Trilogy specification.

Terms in **bold** are defined in the Glossary at the back of this book. At the end of each chapter, you are signposted to pages in *Collins AQA GCSE Combined Science Revision Guide* (ISBN 9780008160791) for more information on the topics covered. The same page references apply to *Collins AQA GCSE Combined Science All-in-One Revision & Practice* (ISBN 9780008160869).

AQA GCSE Science Exams

You will sit six exams, each of 1 hour 15 minutes duration.

In the final year of your GCSE course, your school will choose to enter you for either the Higher or Foundation Tier exam. If you are not sure which exam tier you have been entered for, talk to your science teachers.

The information in the table below is the same whichever tier you sit.

	Topics covered	Exam marks	% of overall grade	Types of question
Biology Paper 1	1. Cell biology 2. Organisation 3. Infection and response 4. Bioenergetics	70	16.7%	Multiple choice Structured Closed short response Open response
Biology Paper 2	5. Homeostasis and response 6. Inheritance, variation and evolution 7. Ecology	70	16.7%	Multiple choice Structured Closed short response Open response
Chemistry Paper 1	8. Atomic structure and the periodic table 9. Bonding, structure, and the properties of matter 10. Quantitative chemistry 11. Chemical changes 12. Energy changes	70	16.7%	Multiple choice Structured Closed short response Open response
Chemistry Paper 2	13. The rate and extent of chemical change 14. Organic chemistry 15. Chemical analysis 16. Chemistry of the atmosphere 17. Using resources	70	16.7%	Multiple choice Structured Closed short response Open response
Physics Paper 1	18. Energy 19. Electricity 20. Particle model of matter 21. Atomic structure	70	16.7%	Multiple choice Structured Closed short response Open response
Physics Paper 2	22. Forces 23. Waves 24. Magnetism and electromagnetism	70	16.7%	Multiple choice Structured Closed short response Open response

Grading and Certification

The qualification will be graded on a 17-point scale: 1–1 to 9–9, where 9–9 is the highest grade. The grade is determined by your total mark out of the 420 marks available from the six papers.

If you are taking the Foundation Tier exam, then you will be awarded a grade within the range of 1–1 to 5–5. If you fail to reach the minimum standard for grade 1–1, you will be recorded as U (unclassified) and will not receive a qualification certificate.

Introduction

If you are taking the Higher Tier exam, you will be awarded a grade within the range 4–4 to 9–9. If you are sitting the Higher Tier exam and you narrowly fail to achieve grade 4–4, you will be awarded a 4–3. If you fail to reach the minimum standard for the allowed grade 4–3, you will be recorded as U (unclassified) and will not receive a qualification certificate.

Assessment Objectives

There are three assessment objectives (AOs) and the six exams will test these three different areas.

Assessment objective	Percentage of exam	Requirements
AO1 Demonstrate knowledge and understanding	40%	Demonstrate knowledge and understanding of: • scientific ideas • scientific techniques and procedures.
AO2 Apply knowledge and understanding	40%	Apply knowledge and understanding of: • scientific ideas • scientific enquiry • scientific techniques and procedures.
AO3 Analyse information and ideas	20%	Analyse information and ideas to: • interpret and evaluate • make judgements and draw conclusions • develop and improve experimental procedures.

Working Scientifically

The study of science involves a lot of facts, theories and explanations but you will also gain an understanding of the scientific process by thinking, discussing and reading about what scientists do. You will look at the contribution that some scientists have made to the world of science and how ideas developed over time using the scientific process. You will also be encouraged to appreciate how scientists communicate and check their ideas through scientific publications, and how ideas find practical and technological applications in everyday life.

Required Practical Activities

Science is a practical subject. Many of the facts and ideas in science are derived from experiments, and theories can be proven or improved by experimental work.

The specification requires that you carry out practical work in 21 specific areas, although you may well do more than this number of practical experiments during your course. Approximately 15% of marks in the exams will be based on the understanding that you have carried out these required practical activities. Questions will draw on the knowledge and understanding gained from having completed these practical activities.

There are many examples of questions based on the required practical activities throughout this book. You can find further information about the required practical activities in the AQA Combined Science: Trilogy specification.

Mathematical Requirements

Some topics in science are quantitative in nature, i.e. they involve the use of numbers to find the answers to questions. There are also some formulae that you are required to know and these are shown on pages 300–303. Note that you are not given a formula sheet in the exam and therefore you should learn these formulae.

There are also many mathematical skills that you will need to be confident with, such as rearranging equations and calculating percentages. Whilst the questions in this book cover the mathematical skills required, you can find further information about the mathematical skills in the AQA specification.

Success in Science

Science can sometimes be difficult! There are lots of facts to learn, theories to understand and explain, and you are expected to apply your knowledge in different contexts. Science is made all the more challenging because you can't always see what is happening with the naked eye and this can make it difficult to try and understand what is going on. The use of models in science can help us to visualise and contextualise things that we can't see or are difficult to understand.

Practising questions such as those given in this book will help you to assess your knowledge and understanding. It will also help you to learn how to answer exam questions successfully. However, for factual information and to aid your understanding of the subject, use this book in conjunction with *Collins AQA GCSE Combined Science Revision Guide* (ISBN 9780008160791) or *Collins AQA GCSE Combined Science All-in-One Revision & Practice* (ISBN 9780008160869).

Command Words

Command words are specific words in questions that tell you what is expected in your answer. Command words can help you decide how to answer questions, how much detail to give and whether you are expected to recall information or give more explanation.

As you work through this book identify the command word in the question. If you are not sure of what the question is asking you to do then refer back to the list below.

Calculate	use numbers given in the question to work out the answer
Choose	select from a range of alternatives
Compare	this requires the student to describe the similarities and / or differences between things, not just write about one
Complete	answers should be written in the space provided, for example on a diagram, in spaces in a sentence, or in a table
Define	specify the meaning of something
Describe	students may be asked to recall some facts, events or processes in an accurate way
Design	set out how something will be done
Determine	use given data or information to obtain an answer
Draw	produce, or add to, a diagram
Estimate	assign an approximate value
Evaluate	students should use the information supplied, as well as their knowledge and understanding, to consider evidence for and against
Explain	make something clear, or state the reasons for something happening
Give	only a short answer is required, not an explanation or a description
Identify	name or otherwise characterise
Justify	use evidence from the information supplied to support an answer
Label	provide appropriate names on a diagram
Measure	find an item of data for a given quantity
Name	only a short answer is required, not an explanation or a description; often can be answered with a single word, phrase or sentence
Plan	write a method
Plot	mark on a graph using data given
Predict	give a plausible outcome
Show	provide structured evidence to reach a conclusion
Sketch	draw approximately
Suggest	this term is used in questions where students need to apply their knowledge and understanding to a new situation
Use	the answer must be based on the information given in the question. Unless the information given in the question is used, no marks can be given. In some cases students might be asked to use their own knowledge and understanding
Write	only a short answer is required, not an explanation or a description

Cell Biology

Animal and Plant Cells

When answering exam questions about different types of cells and the processes they go through, certain technical terms need to be learned and used correctly.

The following multiple-choice question illustrates how learning biological terms can result in achieving easy marks.

Example

Which of the following words best describes a bacterial cell? Tick **one** box.   *(1 mark)*

Prokaryotic

Multicellular

Eukaryotic

Undifferentiated

Prokaryotic ✓

You can quickly rule out 'multicellular' as, clearly, bacteria consist of one cell only (unicellular). More problematic is 'undifferentiated' but once you understand that the process of differentiation only occurs in multicellular organisms then this is also rejected for bacteria. The final two terms relate to the nature of cells themselves. Of the two, 'prokaryotic' is most appropriate because (among other things), prokaryotes have their DNA in the form of a loop (rather than a nucleus) and often contain plasmids. Of course, you could simply learn by rote that bacteria are prokaryotic, but if you have a thorough understanding of these terms you can apply the knowledge to more complex and extended questions.

Most students can recognise the differences between animal and plant cells from diagrams and pictures, but often the key differences are misunderstood and cause problems when students are presented with unusual animal and plant cells.

Example

This is a diagram showing two guard cells, which are found in the leaves of plants.

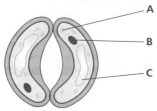

Cell Biology

a) i) Name structure A. *(1 mark)*

> A is a chloroplast ✓.

ii) Explain why this structure would not be found within an animal cell. *(2 marks)*

> It would not be found in an animal cell because chloroplasts are sites of photosynthesis ✓, and animal cells do not photosynthesise ✓.

b) Which structure, B or C, would be found in almost all animal and plant cells? What is its function? *(2 marks)*

> B (nucleus). It contains genetic material / chromosomes and controls other cell processes ✓ involved in cell division ✓.

A common type of question involves an understanding of scale and the ability to carry out simple magnification calculations.

Example

Darren is looking at some cheek cells under the microscope using high power. He clearly sees the nucleus, cytoplasm and cell membrane of the cheek cells.
He wishes he could see mitochondria and ribosomes.

Give **two** reasons why he cannot see these structures. *(2 marks)*

> ✓ ✓ Any two from:
> the organelles are too small to see
> the microscope doesn't have a high enough resolving power
> organelles need to be stained.

c) HT Using a specially fitted camera, Darren takes a picture of the cells he sees.

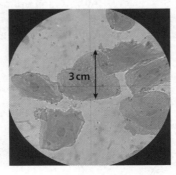

3 cm

Not to scale

He measures the diameter of one cell on his photograph by drawing a line and using a ruler. The line is shown on the picture. It measures 3 cm.

If the microscope magnifies the image 400 times, calculate the actual size of the cell in μm.

You can use the following formula:

$$\text{magnification} = \frac{\text{size of image}}{\text{size of real object}}$$

(3 marks)

size of real object $= \frac{3}{400}$ ✓

$= 0.0075\,cm$ ✓

$= 75\,\mu m$ ✓

Inserting the correct answer usually gains you full marks for this type of question, but you should always show your working because if you arrive at the wrong answer (perhaps by not pressing the correct key on your calculator or making a rounding error) then you can gain credit for showing how you intended to calculate the answer. To arrive at the answer you have to rearrange the formula to make 'size of real object' the subject of the formula,

i.e. size of real object $= \dfrac{\text{size of image}}{\text{magnification}}$

The example also includes an extra level of difficulty in that you have to show your answer as micrometres. You therefore need to know how many micrometres there are in a centimetre and convert accordingly. In this case, there are 10 000 micrometres in a centimetre so multiply the answer (0.0075) by 10 000.

Organisation

Levels of organisation within multicellular organisms refer to how cells are organised into tissues, tissues into organs and organs into organ systems. Simple exam questions are often aimed at testing your ability to appreciate order of magnitude and scale.

Example

Which of the following structures are in the correct size order, starting with the smallest? Tick **one** box.

(1 mark)

Tissue, organ, system, cell ☐	Cell, organ, system, tissue ☐
Cell, tissue, organ, system ☐	Tissue, system, cell, organ ☐

Cell, tissue, organ, system ✓

The correct order is usually quite easy to identify, but make sure you start with the correct item. For example, in this case, the question asks for smallest first. An easy mark can be lost by overlooking this detail.

Differentiation

Differentiation is related to cell specialisation in that multicellular organisms have a wide variety of different forms of cells, all arising from the same basic type. The process by which this is achieved is called differentiation.

Example

Diagram A shows a fertilised egg or **zygote**. All cells in the body arise from this one type, including cell B.

Zygote cell

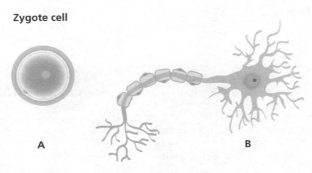

A   B

a) Name cell B.   *(1 mark)*

> (Motor) *neurone / nerve cell* ✓

b) Explain how **one** adaptation of cell B allows it to perform its function efficiently.
(2 marks)

> *Dendrites* ✓ *allow connections to many other nerve cells* ✓.
> Or
> *(Long) axon* ✓ *allows transmission of nerve impulse over long distances* ✓.
> Or
> *Myelin sheath* ✓, *for insulation / to speed up nervous transmission* ✓.

To ensure that full credit is given, the proper technical name should be given to the structure (dendrites or axon) and full detail should be given to the adaptation (many other nerve cells / long distances). For example, the word 'many' is crucial in the example of the dendrites because a connection could still be made with only a single nerve fibre or axon. The point is that many connections allow control over responses through the synapses. They also allow a greater complexity (although these two reasons are not asked for in the question).

In the case of the 'axon' response, the emphasis here is on carrying the nerve impulse over a long distance – not just allowing the impulse to be carried.

Stem cells refer to undifferentiated cells present in the embryo, some of which remain until adulthood. Stem cells have the potential to become almost any type of cell. Most exam questions focus on the use of these cells in medicine to treat a range of conditions.

Example

New research has taken place that allows a fully formed kidney to be developed from foetal cells obtained from amniotic fluid in the uterus (womb). In theory, doctors could collect amniotic fluid from around the baby, then store it for use in the baby's later life.

a) Describe how such cells could be useful in later life. *(2 marks)*

✓ ✓ Any two from:

for therapeutic cloning

for producing cells with the same genes as the patient, e.g. for treating diabetes

Or

for transplant organs

stem cells can be used to produce new organs that won't be rejected by the recipient, e.g. kidney.

The important principles to stress here are that the stem cells can become any type of cell, and that they will have the same genes as the person in question.

b) Explain why the foetal cells are more useful than cells extracted from the individual several years after birth. *(2 marks)*

Cells in later life are already differentiated / become fixed ✓ and cannot become other types of cells ✓.

c) Other uses of stem cells often require research using extracts from embryos. Explain **one** reason why some people might object to this type of research. *(2 marks)*

✓ ✓ Any two from:

an embryo is seen by some people as a living / sentient being

and it would be unethical to experiment on its cells

Or

some stem cells might act as a reservoir of cancer cells

that could then spread to other parts of the body.

There are **two** marks for this question, so it is important to give **two** scientific points to gain full credit.

Cell Division

Mitosis and **meiosis** are both forms of cell division and are often confused in terms of the process involved and the biological function of each one.

Example

Diagram A shows a vertical section through an onion root tip as seen under the light microscope at high power. Diagram B shows some of these cells at an even higher magnification.

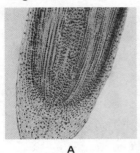

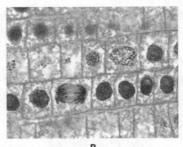

A   **B**

a)   Name the type of cell division taking place in the root tip.   *(1 mark)*

Mitosis ✓   The process can be identified because meiosis only occurs in reproductive tissue such as a testis or ovary. As an onion root tip is neither, the only option is mitosis. You might also deduce that root tips are rapidly growing meristems and growth is achieved by mitosis.

The graph below shows the masses of DNA found in the root tip cells.

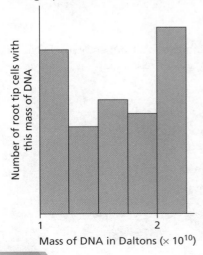

Number of root tip cells with this mass of DNA

1   2

Mass of DNA in Daltons ($\times 10^{10}$)

b) Explain why some cells have a mass of 1×10^{10} Daltons while others weigh 2×10^{10} Daltons. *(2 marks)*

> *Cells with the lower mass have the full chromosome number / have not had their DNA replicated yet (as part of mitosis)* ✓. *Cells with the higher mass have completed DNA replication / chromosome number has doubled (prior to the cell dividing)* ✓.

> You will be given credit for showing that you understand the process of mitosis: DNA is doubled up or replicated *then* the cell divides – which halves the mass again. The graph shows cells in both states as mitosis does not happen simultaneously between cells.

c) State another reason, apart from growth, why a cell might carry out this process of cell division. *(1 mark)*

> **Asexual reproduction** / *repair / replacement of old cells* ✓.

Example

Cells in reproductive organs divide by meiosis to form **gametes**. Describe what happens during this process and explain why changes in the number of chromosomes occur. *(4 marks)*

> This is an example of an extended answer question and examiners will be looking for **four** definite scientific points. You are likely to gain maximum credit if your sentences make sense when read together. Marks are often lost due to the explanation being difficult to follow – even though the student thinks they have correctly included the scientific information!

> ✓ ✓ ✓ Any three from:
>
> *copies of the genetic information are made*
>
> *the cell divides twice to form four gametes*
>
> *each daughter cell has a single set of chromosomes*
>
> *all gametes are genetically different from each other.*
>
> ✓ At least one explanation from:
>
> *gametes join at fertilisation / halve during meiosis to maintain the normal number of chromosomes*
>
> *meiosis / chromosome division allows* **variation** *to occur.*

Diffusion

There are three physical processes you need to be aware of that explain how materials enter and leave cells.

The first of these is diffusion. Questions will probe your understanding of the factors that affect the rate of diffusion, and adaptations found in organisms to increase this rate.

HT Example

Here is some data about a selection of different-sized organisms.

Organism	Surface area / m^2	Volume / m^3	Surface area (SA : vol ratio)
E. coli bacterium	6×10^{-12}	10^{-18}	
Malarial parasite	6×10^{-8}	10^{-12}	60 000:1
Honey bee	6×10^{-4}	10^{-6}	600:1
Roe deer	6×10^{0}	10^{0}	6:1
Sperm whale	6×10^{4}	10^{6}	0.06:1

a) Calculate the surface area : volume ratio of the *E. coli* bacterium. *(2 marks)*

6 000 000:1 ✓ ✓

To work out surface area : volume ratio, it seems a straightforward exercise at first to simply divide the surface area by the volume. However, plugging these values into a calculator will give you a number or decimal. Ratios can be tricky because of the way they are expressed.

There are many ways to tackle the exercise. Here's one: This problem is a little easier than most because you will notice that the ratios in the table are all multiples of 6. If we look at the two numbers, 6×10^{-12} and 10^{-18}, ignoring the scientific notation we see a simple relationship of 6 and 1. We just need to get the orders of magnitude right. If we subtract the indices (10^{-12} minus 10^{-18}) we get 10^{-6}. In other words, the surface area is one million times greater than the volume – hence 6 000 000 to 1.

One more thing – check the units of area and volume. In this case they are expressed in metres (squared and cubed), so we don't have to convert (thankfully – the question is difficult enough as it is – but then, it is Higher Tier!).

b) Using the data, explain why the sperm whale requires lungs and a circulation system to distribute oxygen to its tissues, whereas the *E. coli* bacterium does not. *(3 marks)*

Understanding SA : vol ratios requires you to do a 'flipping' exercise in your mind:

Large animal – small SA : vol ratio

Small animal – large SA : vol ratio

Notice that the question asks you to use the data, so you will not gain full credit unless you refer to the figures and compare them.

✓ ✓ Any two from:

large animals have small SA : vol ratios

passive diffusion is not efficient enough to deliver oxygen to such a large volume of tissue (in the sperm whale)

(therefore) a transport system is needed to transport the oxygen the long distances required.

Or any two from:

small animals have large SA : vol ratios

diffusion is efficient enough to deliver oxygen

because the diffusion path is short (in the bacterium).

And:

comparison of bacterium – 6 000 000:1 compared with sperm whale – 0.06:1 ✓.

c) Use ideas about surface area to explain how mesophyll cells in a plant leaf are adapted to maximise diffusion of water vapour to the environment. *(3 marks)*

Spongy mesophyll cells have air spaces between them ✓, which maximises their surface area ✓. Water evaporates from large surface areas more rapidly ✓.

You need to focus on the spongy mesophyll cells here as the palisade mesophyll cells are adapted for efficient photosynthesis, not evaporation of water / transpiration. Also, do not be drawn into referring to stomata. These control the amount of water vapour lost, which may mean minimising the surface area rather than maximising it.

Cell Biology

Example

Joffrey is conducting an experiment to investigate the effect of surface area : volume ratio on the rate of diffusion.

- He takes some agar blocks containing a green-coloured indicator substance.
- Using a scalpel, he cuts out some jelly cubes with the dimensions shown in this table.

Dimensions of block / mm	SA : vol ratio (expressed as a decimal)	Time for block to completely change colour(s)
10 × 10 × 10	0.6	1400
10 × 10 × 5	0.8	1200
10 × 5 × 2	1.6	550
5 × 2 × 2	2.4	450
2 × 2 × 2	3.0	400

- He fills five boiling tubes with 1 M hydrochloric acid.
- He adds the blocks to the acid in the tubes and starts a stop-clock.
- The acid penetrates the blocks by diffusion and changes the colour of the indicator to red as it does so. Times are taken for each block to completely change colour.
- He records the results shown in the table.

a) Plot a graph of 'time taken for colour change' against 'SA : vol' on the grid. Join the points with an appropriate curve. The axes have been prepared for you.

(3 marks)

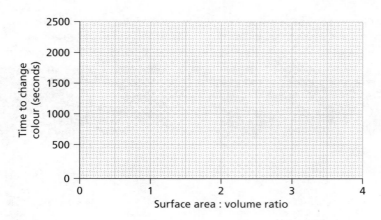

This question provides you with axes and scale – not all questions will. To gain full marks:

- Use a sharp pencil initially (you can rub out any mistakes!). Then go over the plots with a black pen. Circled dots or simple small crosses are acceptable.
- Check the scales – it's easy to misjudge what each small graph square represents. In this case it is:

  vertical (*y*-) axis: one small square represents 50 seconds
  horizontal (*x*-) axis: one small square represents 0.05.
- Use a pencil to produce a smooth curve then go over it in black ink. Be quite deliberate and try to avoid a shaky line (you will be given some leeway by the examiner but you will be penalised for a double line). The line should not extend beyond the data (i.e. past the first or last plots). The data in these questions will usually allow the line to pass through all the points unless there is an anomaly. Some questions ask you to identify an anomalous point (a point that does not fit the curve).

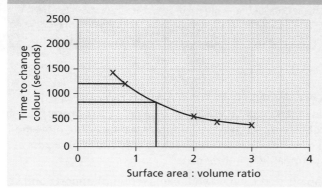

Correct plotting ✓ ✓
(subtract 1 mark for every incorrect plot)

Smooth curve ✓

b) What would be the SA : vol ratio of the block that would completely change colour at 800 seconds? *(1 mark)*

1.35 *(plus or minus 0.05)* ✓

Use a ruler to read off the scales (see answer graph in part **a)**). Again, understanding the scale is crucial. This question requires a high degree of precision (down to two decimal places) although there is a margin for error.

c) Describe the pattern shown by these results. *(2 marks)*

As the SA : vol ratio increases, the time for colour change decreases ✓. The effect decreases markedly after a SA : vol ratio of 2.0 / difference in time for increasing ratios is less ✓.

The first mark is always quite easy to pick up in these 'describing pattern' questions. The second mark is rarely gained because students fail to notice that this is not simply a proportional or inversely proportional relationship (straight line graph). If it was, you should state the straight line relationship.

d) Write down **one** way in which Joffrey could ensure that he obtained valid results in his investigation. Give a reason for your choice. *(2 marks)*

> ✓ ✓ Any two from:
>
> *ensure that all cubes are added at the same time*
>
> *so that each block has the same starting point / times can be fairly compared*
>
> *same volume of hydrochloric acid / agar blocks should be completely covered*
>
> *so that acid can penetrate all blocks equally*
>
> *temperature*
>
> *acid concentration.*

Validity is connected with 'fair testing' and you should concentrate on controlled variables to gain credit. Don't get tied up with accuracy or precision here, e.g. 'cut the blocks carefully'. There are many controlled variables you could choose – always look for the most obvious one(s).

Osmosis

The second passive process you are required to understand is **osmosis**, which is a special case of diffusion. Easy marks can be lost when explaining this process due to not defining terms accurately. Here's a classic definition of osmosis: 'The movement of water from a low solute concentration to a high solute concentration across a partially (or differentially) permeable membrane.'

There are various starting points to setting up your explanation of osmosis, and some will serve you better than others – in fact there is one that could be considered as the 'go to' choice.

Solute concentration – solute refers to the compound dissolved in the water. In biology examples this is usually salt or sugar. Notice that water moves against the solute concentration gradient – or low to high. It is this that confuses students because you are used to understanding diffusion occurring from high to low concentration. But remember, it is water that is moving and not the solute (and it certainly isn't the concentration that moves – as some candidates seem prone to describing!).

Water concentration – to avoid the above confusion, many students lay out their explanation in terms of water concentration as this allows them to equate it with the diffusion model of high to low that they are used to. This method is becoming frowned upon beyond GCSE and, although you will 'get away with it' at GCSE, it's best to opt for the next choice.

Water potential – examiners recommend this approach because it is a measurable parameter that can be easily understood (and marked). To all intents and purposes, you can understand it in the same way as water concentration, but it has a more solid scientific basis. Perhaps it is best to think of it as a 'push' of water as if it was falling from a height.

The next example will illustrate this whilst looking at a particular application of the principle.

Example

Catelyn is investigating how plant cells respond to being surrounded by different concentrations of salt solution. She places rhubarb epidermal cells into 1.0 M salt solution and then observes them under the microscope. This is what she sees:

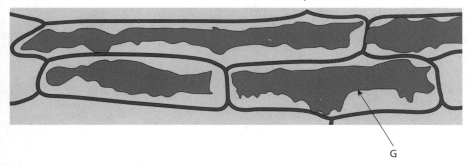

G

a) Name structure G.   *(1 mark)*

Cytoplasm ✓

b) Describe and explain the appearance of the rhubarb cells.   *(3 marks)*

✓ ✓ ✓ Any three from:

cytoplasm / protoplast has pulled away from the cell wall

this is **plasmolysis**

water has left the cells

by osmosis

water potential outside of cell is less than that inside / solute concentration is greater outside of cell / (water concentration is less outside of the cell)

water moves down the water potential gradient / against the solute concentration gradient.

turgor pressure on cell wall is reduced.

You see here that credit is given for using correct terminology like 'plasmolysed'. Always use the correct technical terms where possible. Also notice that credit is given for using a water concentration argument, but be aware that this might not be accepted in the future.

c) Catelyn removes the cells and places them in distilled water. Describe how the cells would change in appearance if she observed them under the microscope again. (*1 mark*)

> ✓ Accept any one from:
>
> *cytoplasm would be pushed up against cell wall / no gap between cell wall and cytoplasm*
>
> *larger vacuole / vacuole swells or increases in size.*

Active Transport

Unlike diffusion and osmosis, **active transport** requires energy to occur. You may be given situations in both animal and plant systems where active transport occurs. You need to be able to identify the features of active transport and apply them to the example. These features are:

- movement of particles / molecules / ions against a concentration gradient
- requires release of energy (from respiration).

Example

The diagram below shows a microscopic view of a root hair cell absorbing ions from the soil water that surrounds it. Explain how the ions can enter despite the fact that this is against a concentration gradient. (*2 marks*)

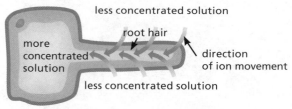

less concentrated solution

root hair

more concentrated solution

direction of ion movement

less concentrated solution

> ✓ ✓ Any two reasons from:
>
> *ions enter by **active transport***
>
> *requiring release of energy*
>
> *from **respiration***
>
> *via protein carriers in the cell membrane.*

Reference to protein carriers is no longer required for the AQA specification, but credit is usually given for this level of detail at GCSE. Once again, use of technical terms like 'active transport' gains easy marks.

For more on the topics covered in this chapter, see pages 16–23 of the *Collins AQA GCSE Combined Science Revision Guide*.

The Digestive System

You will have learned much about the anatomy of the digestive system at Key Stage 3 and will be required to have a thorough working knowledge of this for GCSE exams. Questions often focus on the connection between named organs and tissues, together with their role in both physical and chemical digestion.

Example

The diagram shows some organs in the human digestive system.

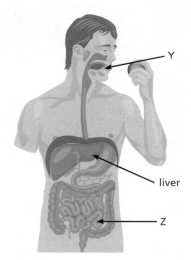

a) A process of absorption occurs in organ Z.

i) Name organ Z. *(1 mark)*

> Small intestine / ileum ✓

ii) In order for food to be absorbed in organ Z, it has to be digested first. Explain why this has to occur. *(2 marks)*

> To break large molecules down into smaller ones ✓, so they are small enough to pass across the gut wall ✓.

iii) The same type of enzyme (**amylase**) is produced in both organs Z and Y and acts on starch. Complete the word equation for this reaction.

starch $\xrightarrow{\text{amylase}}$ _____ *(1 mark)*

sugar / maltose ✓

AQA require you to know the enzyme groups: amylase, proteases and lipases. Glucose would be incorrect.

iv) A student carries out this reaction in a test tube using artificial amylase and a suspension of starch. State the reagent they would use to test for the product, and the practical steps needed to reveal a positive result. *(3 marks)*

This question relates to a required practical you will have carried out during your course. You will have used various reagents to test for a range of carbohydrates, lipids and proteins. It is important that you remember the names of these reagents, the experimental steps taken to use them and the result you would observe during a positive test.

Benedict's reagent ✓

✓ ✓ Any two from:

Add an equal volume / excess of reagent to the tube

heat in a water bath set at a temperature of at least 80°C

colour should change to green / orange / brick-red if reducing sugar (product of digestion) is present.

b) i) The liver is labelled on the diagram and is connected to an organ that stores bile. Name this organ. *(1 mark)*

Gall bladder ✓

ii) Bile emulsifies fat into droplets. Explain how this helps aid chemical digestion. *(2 marks)*

Increases the surface area of the lipid ✓, *which enables the enzyme lipase to attack / react with it more efficiently* ✓.

Many students can gain one mark on this type of 'explain' question, but few gain the full credit. When you see multiple marks available like this, ensure that you include sufficient detail / scientific points. It's worth checking this when you review your paper in the last ten minutes of the exam.

Enzymes

A number of common mistakes occur when students approach questions about enzymes.

- They are **molecules** (quite large ones), not cells and are not to be confused with hormones or antibodies.

- They are not alive and therefore they cannot be killed. You will lose marks if you make this mistake. They can, however, be **denatured**, which occurs when they are exposed to excessive heat or pH extremes.

- They break food molecules down without being chemically changed themselves.

Example

In an experiment to investigate the enzyme catalase, potato extract was added to a solution of hydrogen peroxide. The catalase in the potato catalysed the decomposition of the hydrogen peroxide and produced oxygen bubbles. The experiment was carried out at different temperatures and the results recorded in the table below.

Temperature (°C)	1	10	20	30	40	50	60	70	80
Number of bubbles produced in one minute	0	10	24	40	48	38	8	0	0

a) i) Plot a graph of these results on graph paper and join the points with a smooth curve. *(3 marks)*

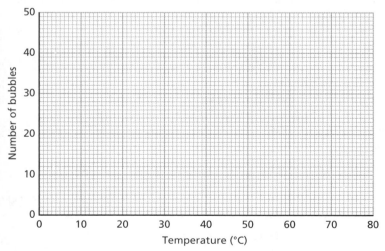

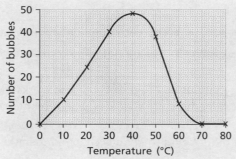

2 marks for correct plotting ✓ ✓
Subtract 1 mark for every incorrect plot.

1 mark for smooth curve ✓

ii) Describe how the number of bubbles produced varies with the temperature of the reacting mixture. *(2 marks)*

> *The rate of bubbles produced increases until it reaches an optimum / maximum* ✓, *then it decreases rapidly, producing no bubbles at 70°C* ✓.

iii) Using the graph, estimate the optimum temperature for catalase to work at. *(1 mark)*

40°C ✓   The optimum temperature is obtained from reading off vertically from where the graph peaks. In this case the peak coincides with one of the plotted points but sometimes your curve needs to be extrapolated between two data points. As in all cases there will be a margin of error with the graph – usually equal to one small square (equivalent to 1°C in this example).

b) i) In 1890, Otto Fischer put forward a theory called 'lock and key', which provided a mechanism to explain how enzymes work. Explain, using lock and key theory, how catalase can break down hydrogen peroxide molecules. *(4 marks)*

> ✓ ✓ ✓ ✓ Any four from:
> *catalase / enzyme has a specific shape that will fit only the peroxide molecule (like a lock and key)*
> *this part of the enzyme molecule is called the* **active site**
> *the enzyme puts a physical strain on the bonds in hydrogen peroxide*
> *the peroxide molecule is broken apart*
> *as the products no longer fit the active site.*

ii) In 1958, Daniel Koshland updated Fischer's ideas and then published his findings. Apart from modifying existing theories, explain why scientists publish their work in scientific journals. *(1 mark)*

> ✓ Any one from:
> *to allow other scientists to test / verify their work*
> *to share their ideas so they can be used for the benefit of all*
> *to ensure that false claims are not made / allow peer review.*

Breathing

You will need to recall the major organs and tissues within the human breathing system and be able to label these if provided with a diagram. It is important to distinguish between breathing (or ventilation) and respiration. Breathing is the process by which air enters the lungs using contractions of the diaphragm and intercostal muscles. Respiration refers to energy release in cells and is dealt with in the 'Bioenergetics' section.

Example

The lungs are organs of gaseous exchange. Explain how alveoli in the lungs are adapted to absorbing maximum oxygen into the bloodstream. *(4 marks)*

Alveoli are present in large numbers and so provide a large surface area for the exchange of gases ✓, the thin alveolar wall reduces the diffusion path ✓, the extensive capillary network maximises absorption of gases ✓, moist lining / epithelium of alveolus allows gases to dissolve / speeds up diffusion ✓.

Take care in describing adaptations. You need to ensure that you link the feature to a reason and say why the structure represents a successful or efficient adaptation.

Questions about breathing are often connected to lung diseases or to smoking or cancer. Be prepared to relate your knowledge of the breathing system to data analysis.

Example

Scientists are studying the performance of pearl divers living on a Japanese island. They have taken measurements of the lungs of five 20–30 year-olds and timed how long they can stay underwater. The scientists also measured recovery time for the divers' breathing rates after a dive. The data is shown in the table below.

Vital capacity / litres	Max. time under water / mins	Time for breathing rate recovery / mins
3.5	2.5	3.2
4.0	2.7	2.9
4.3	2.8	3.5
4.5	2.9	2.8
4.6	3.0	2.5

a) Using your knowledge of the alveoli and diffusion, explain why pearl divers cannot stay underwater for longer than three minutes. *(2 marks)*

> ✓ ✓ Any two from:
>
> concentration of oxygen in alveolus falls
>
> as it is not being replenished through breathing
>
> and oxygen is removed to the bloodstream
>
> concentration gradient falls.

b) Vital capacity is the maximum volume of air that the lungs can contain. One of the team of scientists suggests that having a larger vital capacity allows a diver to stay underwater for longer. How well does the evidence support this conclusion? *(2 marks)*

> Agree (no mark): because as vital capacity increased the time underwater also increased ✓.
>
> Disagree (no mark): any one from: only five subjects / sample size too small; need to include wider range of vital capacities ✓.

Notice that your decision to agree or disagree carries no marks as there are valid points on both sides of the argument. Also notice that to gain full marks you need to look at both possibilities and show an awareness of how data needs to be treated with caution. This is an element of working scientifically that appears throughout your exams.

c) The team decides that the vital capacity and breathing recovery rate is inconclusive. Suggest how they could obtain more valid data. *(2 marks)*

> ✓ ✓ Any two from:
>
> test a wider range of subjects
>
> need to take account of the diver's gender
>
> standardise the ages (keep them the same)
>
> standardise the fitness levels.

The Heart

Humans have a **double circulatory system** that allows blood to remain at a high pressure, as it needs to travel long distances around the body. This means that for every circuit of the body, the blood receives a boost two times – once after it has passed through the lungs, and once again when it has returned from the rest of the body. This idea is often tested in exams and could be coupled with questions regarding pressure changes in the heart.

Example

This table shows pressure changes that occur in the left ventricle of the heart.

Time (s)	Pressure (kilopascals)
0.1	1
0.2	8
0.3	15
0.4	15
0.5	3

a) Describe the pattern shown by this data. *(2 marks)*

✓ ✓ Any two from:

pressure increases dramatically with time
until 0.3 / 0.4 seconds
then decreases (after 0.4 seconds).

Here is another example where full credit is gained for seeing all aspects of the pattern and identifying where changes occur – this is usually best achieved by quoting data.

b) At what time will most blood be present in the ventricle? *(1 mark)*

0.1 seconds ✓   Most blood will be present in the ventricle when it is at its most relaxed (ventricular diastole), which is just before contraction. In other words, when the pressure is least.

You need to be able to calculate blood flow rates. Interpreting this information and applying it to situations that might affect the heart are dealt with in the next example.

Example

Rate of blood flow can be calculated in a measure called **stroke volume**. This is the volume of blood pumped out of the heart every beat. The following formula is used:

$$\text{stroke volume (in litres)} = \frac{\text{volume of blood pumped by the heart per minute}}{\text{heart rate}}$$

a) Calculate the stroke volume of an athlete whose heart pumps 7.15 litres per minute at a heart rate of 65 beats per minute. *(2 marks)*

0.11 litres per beat ✓ ✓
If an incorrect answer is given, the correct working (7.15 ÷ 65) would still gain 1 mark.

This again illustrates the importance of showing full working in these calculations as credit can still be gained even if you arrive at the wrong answer.

b) In a condition called dilated cardiomyopathy, or DCM, stroke volume is lowered as the ventricles fail to contract properly. The diagrams below show a healthy heart and a heart with DCM.

Healthy heart   DCM heart

A

i) Name the chamber labelled **A** in the diagram. *(1 mark)*

> Right atrium / auricle ✓

ii) In the wall of this chamber there is specialised nervous tissue that keeps the heart beating in a controlled way. What is it called? *(1 mark)*

> Pacemaker ✓

iii) The data in the table below shows stroke volumes from healthy people and those with DCM.

	Normal heart	DCM heart
Stroke volume at 9 cm³	80	35

Using the information in the diagrams and the table, explain the effects of DCM on the structure and function of the heart, and suggest how this might affect the health of the patient. *(3 marks)*

✓ ✓ Any two from:

(left) ventricle not able to fully contract

therefore less blood expelled from heart

stroke volume reduced.

✓ And one from:

more strain put on heart muscle

heart beats faster to compensate for lower stroke volume

symptoms might include shortness of breath, swelling of the legs, fatigue, weight gain, fainting, palpitations, dizziness, blood clots in the left ventricle (any one symptom).

Each part of the question needs to be addressed to obtain full marks. Biology questions often involve linking structure to function. In other words, describing a tissue, organ or system and then saying how the job of that structure is affected.

Blood Vessels

Arteries, veins and capillaries are the three main types of vessel you need to be aware of at GCSE.

Example

The diagram shows three kinds of blood vessel found in the human body.

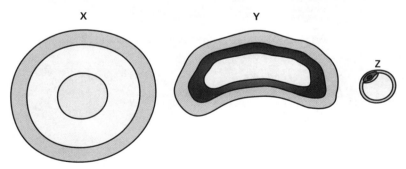

X   Y

Z

Not to scale

a) Name vessels X and Y. *(2 marks)*

X: artery ✓

Y: vein ✓

b) Describe the differences between X and Y, and explain how these differences enable the vessels to carry out their function. *(3 marks)*

> ✓ ✓ ✓ Any three from:
>
> X has narrower lumen / Y has wider lumen
>
> wider lumen in Y gives less resistance to blood flow as pressure is less
>
> X has thicker muscle / elastic wall
>
> to withstand high pressure of blood / to even out pulses / allow smooth flow of blood.

> Notice that in comparison marks you need to be careful about wording (this has been mentioned before). For example, it would not be enough to say that Y has a wide lumen; use of the word 'wider' shows you are comparing it to something else.

c) Vessel Z is a capillary and has a diameter of 0.001 mm or 1 μm. Calculate the cross-sectional area of the vessel using this formula:

cross-sectional area = πr^2 *(2 marks)*

> 0.000000785 mm² or 0.785 μm² ✓ ✓
> If an incorrect answer is given, correct working would still gain 1 mark.

> Radius is half the diameter, therefore 0.0005 mm. Using π as 3.14:
>
> 3.14 × (0.0005 × 0.0005) = 0.000000785 mm²
>
> NB: the calculation is simpler if you use micrometre units.

The graph below shows the pressure changes that occur in the blood vessels lying between the aorta and the vena cava.

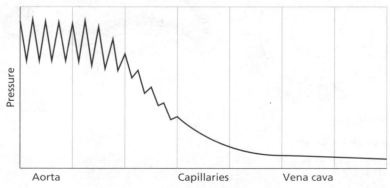

d) Explain why these changes in pressure occur. *(3 marks)*

✓ ✓ ✓ Any three from:

Aorta:
- *high pressure*
- *caused by proximity to heart / muscular force of heart.*

Capillaries:
- *rapid / marked / severe drop in pressure*
- *due to resistance of capillaries with their small cross-sectional area.*

Vena cava:
- *pressure remains low as blood flow rate is slow / vessel is far away from / not yet received force from muscular contractions of heart.*

Blood as a Tissue

Questions on blood as a tissue require you to learn the names of the various components and to relate their structures with their functions.

Example

The diagram below shows some of the components found in blood.

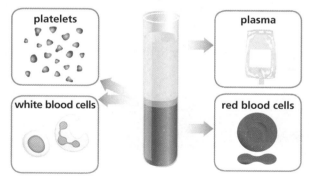

a) State **one** difference in the structure between red and white blood cells. *(1 mark)*

No nucleus in red blood cell / Hb versus no Hb ✓.

b) Why do red blood cells not possess a nucleus? *(1 mark)*

To allow room for molecules of oxygen to be carried / more Hb can fit in ✓.

c) With a diameter of 0.008 mm, a red blood cell is wider than the diameter of a capillary yet can still pass through. Explain how this is possible and why it aids the cell in carrying out its function. *(2 marks)*

> *Red blood cells are flexible / can fold / bend / squeeze through a capillary ✓.*
> *Slower movement of the cell allows time for oxygen to diffuse out into tissues ✓.*

The table shows the concentrations of some materials found in blood plasma.

Material	Concentration in millimoles per litre
Sodium	150
Calcium	0.8
Glucose	5
Protein	70

d) i) The average human body contains 3 L of plasma. How much glucose would this volume contain? (Give your answer in millimoles.) *(1 mark)*

> *15 mmol/L ✓*

ii) This data was obtained from a healthy volunteer. Suggest how the concentrations of these solutes would differ in an untreated diabetic. Give a reason for your answer. *(2 marks)*

> *Glucose concentration higher ✓, due to lack of insulin / inability to absorb glucose from blood ✓.*

This section of your course has been linked with ideas from the 'Homeostasis and response' section. This is a common thing to happen and you should practise linking knowledge between the topics for this reason.

Coronary Heart Disease

The AQA specification covers many non-communicable diseases and their treatments. It is impossible to give examples of every type of question you might encounter, but the following points are quite common:

- Comparing data in populations who suffer from heart disease. This may be shown in charts, graphs and tables. Viewing different presentations of this type of information will help you hone your interpretive skills.

- Describing the advantages and disadvantages of different treatments, for example heart valve replacements versus transplants or mechanical hearts.

- Discussing ethical and economic factors that affect how diseases are treated.

- Understanding the interactions between different factors. For example, heart disease may be affected by hereditary factors, drinking alcohol, diet, smoking, etc. and you should be aware that there is often no single factor that is solely responsible.

Example

a) The diagram below shows the stages of plaque build-up in a coronary artery (atherosclerosis). Such plaques can lead to life-threatening conditions. Use the information in the diagram to explain how plaques can lead to a heart attack. *(4 marks)*

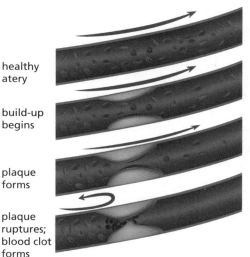

healthy atery

build-up begins

plaque forms

plaque ruptures; blood clot forms

✓ ✓ ✓ ✓ Any four from:

plaque restricts blood flow (in coronary artery)

blood clot more likely to form

oxygen cannot reach heart muscle

heart muscle not able to contract / ventricular fibrillation

heart muscle dies.

The third marking point is only gained if the word 'muscle' is used. Many students lose marks because they refer vaguely to the heart or ventricle.

b) One treatment for coronary heart disease is to introduce a **stent** to the affected artery. Describe how a stent can relieve the problems caused by plaque build-up.

(2 marks)

✓✓ Any two from:

(metal) tube / cylinder inserted into artery

cylinder expanded / inflated to widen artery / lumen

blood flows more freely

more oxygen can reach heart muscle.

c) Prevention of so-called 'lifestyle' diseases is a priority for health services as it saves money in the long term compared with treating the condition once it has happened.

Drugs called **statins** lower cholesterol and some have claimed that they prevent up to 80 000 heart attacks and strokes every year in the UK. The current advice given to GPs is that the benefits of statins far outweigh their risks.

However, there are possible side-effects. One review stated that there is a real risk of myopathy, a neuromuscular disorder that causes muscle damage. One in 10 000 people per year may develop myopathy as a result of taking statins. Another five to 10 people per 10 000 people will have a haemorrhagic stroke, which involves bleeding into the brain.

The graph below shows some data about how long people survive heart disease. It comes from a survey where one group did not take statins whereas the other did.

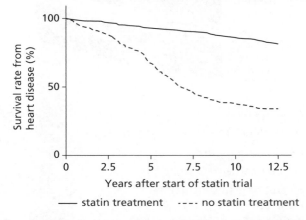

Using the information and the graph, say whether you think statins should be routinely prescribed to everyone over the age of 60 and give reasons for your decision. In your answer you should present arguments for and against prescribing statins.

(4 marks)

✓ ✓ ✓ ✓ Any four arguments from:

For:

statins increase survival rate / life expectancy

people are more likely to have heart disease when they are older (less value in giving statins to younger people)

risks of side-effects are sufficiently low / benefits of taking statins outweigh risks.

Against:

statins can cause life-threatening conditions

numbers may be low but they could be significant and other research needs to be conducted and studied

money could be better spent on other preventative means, e.g. advertising campaigns promoting healthy lifestyles

as healthier lifestyles are risk-free.

Smoking

Example

The table below shows data produced from a Stop Smoking campaign.

Age when men stopped smoking	Risk of cancer (%)
Non-smoker	0.4
30	1.0
40	2.8
50	5.6
60	11.0

a) Predict the risk of lung cancer if a man stopped smoking at 45 years of age. *(1 mark)*

Any answer in the range 2.9–5.5 % ✓.

b) Jorah says that all people who continue to smoke after 60 years should have to pay for NHS treatment arising from smoking-related diseases. Do you agree? Give reasons for your decision. *(2 marks)*

Marks can be gained from either response. ✓ ✓ Any two from the following:

Yes:

risk of having cancer greatly increases between the ages of 50 and 60

lung cancer risk is directly / causally connected to smoking, so the decision to carry on smoking places a burden on the health service / disease is avoidable.

No:

NHS treatment should be offered free whatever the cause of the health problem

difficult to prove that cancer is definitely related to smoking in an individual / cancer might be caused by something else

data only relates to men, and risks may be different for women

unfair to target one lifestyle disease / would have to be applied to alcohol and other drug-related conditions.

Take care to be precise and clear when constructing answers to these types of questions. It is easy to fall into the trap of being vague and giving answers like 'It's their fault that they got the disease' or 'It's unfair to pick on smokers'.

c) Carcinogens in tar are partly responsible for lung cancer. Name **one** other harmful substance found in tobacco and explain how it affects the body. *(2 marks)*

✓ ✓ Either:

carbon monoxide; combines irreversibly with haemoglobin

NB: no marks would be gained for the answer 'makes it harder to breathe'.

Or any one from:

nicotine / highly addictive / raises heart rate.

Example

Emphysema is a lung disease that increases the thickness of the surface of the lungs for gas exchange and reduces the total area available for gas exchange.

Two men did the same amount of exercise. One man was in good health and the other man had emphysema.

The results are shown in the table.

	Man A	Man B
Total air flowing into lungs (dm^3/min)	89.5	38.9
Oxygen entering blood (dm^3/min)	2.5	1.2

a) Which man had more oxygen entering his blood? *(1 mark)*

> Man A ✓

b) Explain why Man B struggled to carry out exercise. *(2 marks)*

> ✓ ✓ Any two from:
>
> less oxygen absorbed into blood
>
> longer diffusion pathway across alveolus wall
>
> insufficient oxygen delivered to muscles to release energy for exercise.

Alcohol

Example

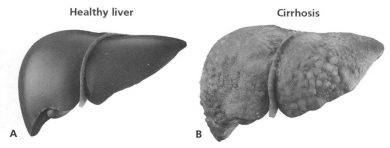

Healthy liver   **Cirrhosis**

A   B

a) Look at the two pictures of the liver. The liver on the left (A) is healthy. The liver on the right (B) is from a person who suffered from a disease called cirrhosis.

Cirrhosis can be caused by excessive alcohol consumption. Explain how over-drinking can lead to cirrhosis of the liver. *(2 marks)*

> Alcohol breakdown produces or releases toxins / poisons ✓.
>
> Hardens / scars the liver tissue ✓.

> You must state that it is the breakdown of the products of alcohol that cause the poisoning of the liver, not the alcohol itself. Note that no credit is given for saying that the liver becomes cirrhosed, as this is already stated in the question. You need to add extra detail to gain the second mark.

A study into alcohol purchasing habits was carried out in a European country. The chart below shows the average number of alcohol units purchased per heavy drinker per year in the study.

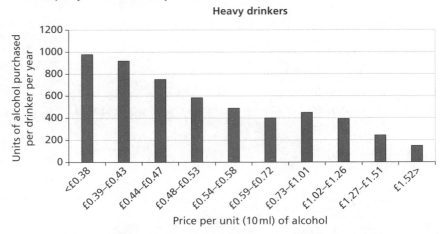

Heavy drinkers

Price per unit (10 ml) of alcohol

b) Calculate the percentage decrease in units purchased between the unit price categories of £0.38 and £1.52. *(2 marks)*

81.6% ✓ ✓

If incorrect the following working can receive 1 mark: $\frac{980 - 180}{980} \times 100$

Where the reading of the graph values is slightly incorrect, answers will be accepted between 80% and 84% as long as they correspond to correct working.

The UK government is thinking of introducing a minimum alcohol unit price of 50p because they say irresponsible drinking costs the taxpayer £21 billion a year, with nearly a million alcohol-related violent crimes and 1.2 million alcohol-related hospital admissions. Others think that it will be unfair to responsible drinkers.

c) Discuss the advantages and disadvantages of introducing such a law. *(3 marks)*

This kind of question involves interpreting data, but also applying it to a situation where economic, health and other issues are involved. As stated many times before, the clarity of the points you make is important.

✓ ✓ ✓ Any three from:

increasing the unit price of alcohol generally reduces the units purchased this may lead to lower alcohol consumption / alcohol-related crime / deaths policy may damage alcoholic drinks industry / the economy / tax receipts may only affect the poor / have little effect on higher incomes / earners more alcohol may be purchased on the 'black market'.

Cancer

The main point to remember about cancer is that it occurs as a result of abnormal cell division. This is why it is connected with mitosis because normal cell division occurs by mitosis. You should also be aware that agents that cause cancer (carcinogens) are many and varied, although a popular group in exam questions are those contained in tar from cigarette smoke.

Example

Scientists investigating the effectiveness of an anti-cancer drug experimented on mice that had their lungs previously injected with a fast-acting carcinogen. One group was given a placebo as a control, while the other was given the anti-cancer drug. There were 10 mice in each group. After five weeks, the mice were killed and their lung tissue examined. The scientists counted the number of tumours they found. The results are shown in the table below.

Number of tumours	Number of mice	
	Group given placebo	Group given anti-cancer drug
0	0	6
1	0	2
2	3	1
3	2	0
4	2	1
5	2	0
6	1	0
7	0	0

The scientists concluded that the drug was effective and that there should be further trials.

a) One group was given a placebo – why?   *(2 marks)*

> *A placebo is a medicine given to a subject without the active ingredient being present ✓, this is to ensure that any effects on the subjects (mice in this case) – beneficial or otherwise – are due to the anti-cancer drug and nothing else ✓.*

b) What evidence is there that the drug is effective? Give a reason why the effectiveness is not necessarily reliable. *(2 marks)*

> Mice given the drug had fewer overall tumours / fewer tumours found on individual mice ✓.
>
> ✓ Any one from:
>
> but no account given to size of tumours
>
> sample size of 10 is quite small
>
> only one trial carried out.

c) Explain how a tumour could develop. *(2 marks)*

> ✓ ✓ Any two from:
>
> mutation in cell
>
> cell divides in an uncontrolled way / fault in the process of mitosis
>
> tumour formed from a group of these mutated cells.

d) The tumours in the mice were **malignant**. What does this mean? *(1 mark)*

> ✓ Any one from:
>
> the cancerous cells invade neighbouring tissues
>
> spread to other parts of the body
>
> secondary tumours formed.

e) The scientists did not recommend proceeding to market the drug immediately. Why not? *(2 marks)*

> ✓ ✓ Any two from:
>
> further research required to see if results are consistently produced
>
> other scientists need to verify / check the work
>
> use larger sample size of mice
>
> clinical trials on humans needed:
>
> to see if the drug is safe / to calculate an effective dose.

f) Some would say this experiment was unethical. Do you agree or disagree? Give a reason for your answer. *(1 mark)*

> This type of question comes up often and you can give a generic answer based on the answers below most of the time. Just be careful not to simply say 'It's cruel'. You might get away with this type of response, but examiners are increasingly looking for greater clarity in these answers.

✓ Any acceptable reason.

Agree:

although the mice suffer / are killed, there is a greater good achieved through developing a drug that relieves suffering in humans / other animals, as the drugs tested on animals often end up being used by vets in the treatment of animals (idea of).

Disagree:

the mice have no choice as subjects in the experiment

there are other methods available for testing the drug / computer simulations / human volunteers.

Plant Tissues and Organs

Be sure to learn the technical names and functions of these plant tissues and organs. Spelling may be important as well, depending on the type of question. It's always good practice to spell correctly anyway.

Example

The diagram below shows a magnified view of the inside of a leaf. Complete the labels. *(3 marks)*

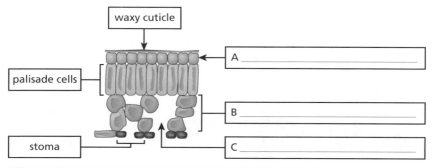

A – (upper) epidermis ✓ *B – spongy layer / mesophyll* ✓ *C – air space* ✓

Xylem and Phloem

Xylem and phloem are often confused in terms of their structure and function. As a quick comparison, always remember the following:

Part of plant	Appearance	Function	How they are adapted to their function
Xylem	Hollow tubes made from dead plant cells (the hollow centre is called a lumen)	Transport water and mineral ions from the roots to the rest of the plant in a process called transpiration	The cellulose cell walls are thickened and strengthened with a waterproof substance called lignin
Phloem	Columns of living cells	Translocate (move) cell sap containing sugars (particularly sucrose) from the leaves to the rest of the plant, where it is either used or stored	Phloem have pores in the end walls so that the cell sap can move from one phloem cell to the next

Example

The diagram shows the structure of phloem in the stem of a common plant. It also shows an aphid (not to scale) inserting its mouthparts into the phloem.

phloem

a) Explain why the aphid inserts its mouthparts into the phloem and not into the xylem. *(2 marks)*

Phloem carries dissolved sugars / sucrose ✓, which the aphid needs for energy / respiration ✓.

This question involves application of knowledge. You may not have heard of aphids before, but if you have revised the fact that the phloem carries sugar, then it isn't hard to deduce why the aphid's mouthparts might extract this rather than water from the xylem.

b) What name is given to the movement of substances around the phloem? *(1 mark)*

Translocation ✓

c) Compare **two** structural differences between phloem and xylem tissue. *(2 marks)*

✓ ✓ Any two from:

xylem tubes are continuous hollow tubes / phloem have end walls / sieve tubes

xylem cells are not living / phloem cells are living

xylem cells are impregnated with lignin / phloem cells are not

phloem has companion cells / xylem does not.

Take care to make your comparisons complete by saying what is present or absent in the phloem and xylem. For example, simply stating that xylem tubes are hollow might not be enough to get the mark if you don't state exactly how phloem cells differ. Some of the comparisons are outside the content required by the specification, but you would gain credit for them anyway.

Transpiration

Transpiration refers specifically to the evaporation of water from the surface of the plant's leaf. More specifically, the water evaporates from the film of water lying over spongy mesophyll cells within the leaf. This is why you need to have a thorough knowledge of plant tissues so that you can relate principles like transpiration to the appropriate structures.

You should also be aware that evaporation of water from leaves drives the whole transpiration stream. This is a continuous process from root to leaves. Many questions are based around the **potometer**, which measures transpiration rates under different conditions. This next example is slightly different but the same principles apply.

Example

In an experiment, a plant biologist carried out an investigation to measure the rate of transpiration in a privet shoot. She set up three tubes like the one in the diagram, measured their mass and exposed them each to different conditions.

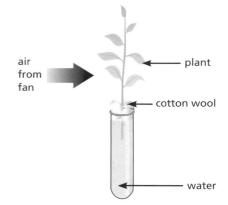

air from fan

plant

cotton wool

water

    A – left to stand in a rack

    B – cold moving air from a fan was blown over it

    C – a radiant heater was placed next to it

Organisation

The tubes were left for six hours and then their masses were re-measured. The biologist recorded the masses in this table.

Tube	A	B	C
Mass at start (g)	41	43	45
Mass after six hours (g)	39	35	37
Mass loss (g)	2	8	5
% mass loss	4.87		11.9

a) Calculate the percentage mass loss in tube B. Show your working. Give your answer to three significant figures. *(2 marks)*

$\frac{8}{43} \times 100 = 18.6\%$

✓ ✓ 2 marks for correct answer but, if incorrect, working gains 1 mark.

b) Which factor increased the rate of transpiration the most? *(1 mark)*

B (cold moving air) ✓

c) Evaporation from the leaves has increased in tubes B and C. Describe how this would affect water in the xylem vessels of the plant. *(1 mark)*

Water column in xylem would move upwards / towards the leaves more quickly ✓.

d) Explain how increasing the movement of air increases transpiration rate. *(2 marks)*

✓ ✓ Any two from:

air movement removes water vapour from air outside of leaf surface

diffusion gradient increased

diffusion of water molecules increased outwards.

Stomata and Root Hair Cells

Example

The diagrams on the next page show a magnified view of the lower leaf surface of a plant. The epidermal cells and stomata are clearly visible.

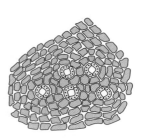

Diagram S

0.5 cm

0.25 cm

Diagram T

a) Calculate the stomatal density in the leaves of plant S. Express your answer in stomata per cm². *(3 marks)*

Area of leaf = 0.5 × 0.25 = 0.125 cm² ✓

16 stomata occupy this area, so the number of stomata occupying

$1 \, cm^2 = \frac{1}{0.125} \times 16$ ✓

= 128 stomata per cm² ✓

The correct answer on its own will yield three marks. But if this is incorrect, you will gain credit for each of the stages of working.

b) Plants lose water vapour through stomata. Explain how the position of stomata and changes in their structure can reduce water loss in plants. *(3 marks)*

Stomata on lower surface of leaf – means they are not as exposed to the heat of sunlight ✓.

Stomata can be closed – by action of guard cells to reduce aperture / exit route of water vapour ✓.

Sunken stomata – increases humidity around stomata and lowers the diffusion gradient ✓.

Notice that the feature has to be linked to the scientific explanation to gain full credit.

c) Plant T has a stomatal density of 10 per cm². Suggest, with a reason, the habitat that plant T might be found in. *(2 marks)*

arid / dry / salt marsh ✓

✓ Any one of:

because lower numbers of stomata suggest the plants struggle to absorb enough water from the soil

fewer stomata means less water loss to counter the fact that little water is absorbed.

For more on the topics covered in this chapter, see pages 24–33 of the *Collins AQA GCSE Combined Science Revision Guide*.

3

Infection and Response

Communicable Diseases

Communicable diseases are those that can be spread via infection. You will need to revise the specific examples given in the specification and know what each one means, together with examples of diseases that are spread by those means.

The following question tests your knowledge of the four types of **pathogen** you need to know about, but you will also need to be aware of the differences between these microorganisms. Of particular note is the difference between bacteria and viruses in terms of size (viruses are much smaller) and the fact that viruses cannot survive for long outside of living cells.

Example

a) Link the type of microorganism to the disease it causes. *(3 marks)*

Bacterium		HIV
Fungus		Malaria
Virus		Salmonella
Protist		Rose black spot

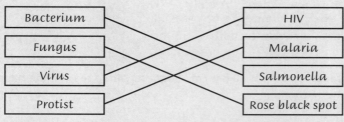

All correct = ✓ ✓ ✓, three correct = ✓ ✓, one or two correct = ✓.

b) The graph below shows the number of reported cases of salmonella per 100 000 of the population over the period of one year.

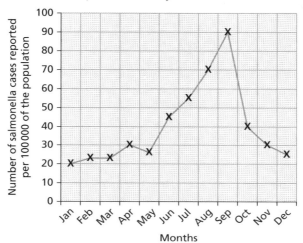

Months

Salmonella is spread when food is undercooked or stored at the wrong temperature. Use this information and the graph to describe and explain the trend shown. *(3 marks)*

✓ ✓ ✓ Any three from:

number of cases increase dramatically in Summer months / June to September e.g. from 44 per 100 000 to 88 per 100 000

warmer temperatures in Summer encourage growth of salmonella bacteria / vice-versa for Winter months

untended food warms up more rapidly in the summer months, encouraging bacterial growth.

Viral Infections

Example

John has caught measles and has been confined to bed for several days. John's mother was immunised against measles as a child via the MMR (measles, mumps and rubella) vaccine. She did not have John vaccinated as she was worried about reports in the media concerning a possible link between the MMR vaccine and autism.

a) How is John likely to have become infected with measles? *(1 mark)*

*Via **droplet infection** / virus carried in the air when a person with the disease coughs, sneezes or breathes ✓.*

b) State **one** symptom John may suffer as a result of the measles. *(1 mark)*

> ✓ Any one from:
>
> fever   loss of sight
>
> red skin rash   encephalitis
>
> in some cases – fatal / death

c) The measles virus is a pathogen. What **two** ways do pathogens produce the symptoms that John experiences? *(2 marks)*

> Cell damage ✓
>
> Production of toxins ✓

d) The study that seemed to support a link between autism and the MMR vaccine was later discredited, even though it was originally published in a reputable scientific journal. The reputable journal later retracted the paper. Suggest how scientists might have discovered that the claim made about the link was false.
>
> *(1 mark)*

> ✓ Any one from:
>
> by studying the data again
>
> by producing new evidence / looking at other studies
>
> by checking the data analysis / conclusions drawn.

The other example from the specification of a viral infection you need to know about is HIV. This virus is sexually transmitted. It is important to realise that HIV refers to the virus, whereas AIDS is a description of the 'full-blown' symptoms resulting from untreated HIV infection. The following example brings in ideas about the immune system, which is typical of 'infection'-based questions.

Example

HT The photograph shows the structure of the human immunodeficiency virus (HIV). For decades HIV has spread throughout the world, especially in developing countries.

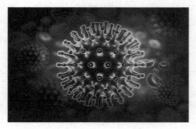

A vaccine is now being developed that shows promising results. It works by mimicking the shapes and structures of HIV proteins. Scientists hope the immune system may be 'educated' to attack the real virus. A specially designed adenovirus shell can protect the vaccine genes until they are in a cell that can produce the vaccine protein.

Using your knowledge of the immune response and immunological memory, describe and explain how antibodies can be produced against the HIV virus.   *(6 marks)*

✓ ✓ ✓ ✓ ✓ ✓ This is a model answer that would score the full 6 marks.

HIV proteins can be triggered and manufactured in existing human cells. The genes that code for the viral proteins are injected into the bloodstream. An adenovirus shell prevents them from being destroyed by the body's general defences. Once inside a cell, the genes instruct it to produce viral proteins that are presented at the cell surface membrane. The body's lymphocytes then recognise these antigens and produce antibodies against them. Memory cells sensitive to the viral proteins are stored in case the body is exposed to the antigens again.

This is an example of a 'level of response' type question. It is not marked in the usual way. Rather, the examiner will weigh up the scientific points you have made and how coherently you have presented them. This is why the answer is given in the form above instead of a series of marking points. Generally speaking, the levels are awarded as follows.

Level 3: A clear, logical and coherent answer, with no wasted words or excess detail. The student understands the process and links this to reasons for any appropriate experimental approaches. **5–6 marks**

Level 2: A partial answer with errors and ineffective reasoning or linkage. **3–4 marks**

Level 1: One or two relevant points but little linkage of points or logical reasoning. **1–2 marks**

Plant Infections

TMV and rose black spot are two pathogens the AQA specification requires you to understand. TMV is a viral pathogen that affects a variety of plants including tomatoes, while rose black spot obviously affects roses. This next example looks at both diseases and brings in ideas about disease control and the use of inorganic chemicals on the land.

Infection and Response

Example

Rose black spot again. Fungicide is the only solution. It's straightforward and easy to apply.

It's better to avoid chemicals if you can – they're bad for the environment.

Dani   Ramsay

Dani and Ramsay both own rose gardens and want to produce high quality flowers for showing at competitions. Both their gardens have been affected by rose black spot in the past, but they differ in their approach to control.

a) Describe the symptoms and cause of rose black spot and discuss the options that Ramsay and Dani might consider. *(4 marks)*

> *Caused by a fungus ✓, which produces purple or black spots on leaves, which often turn yellow and drop early ✓. Treatment with fungicide – will kill the fungus on the leaves but disease may remain in soil or on other leaves / plants and cause re-infection ✓, removal of leaves / ensuring that plants are healthy / well-watered / have access to nutrients is more time-consuming but has a preventative effect too ✓.*

The mark scheme suggests how ideas can be expressed in an answer, but the examiner is interested in the idea itself, and as long as it is explained clearly you will gain credit.

b) Another plant pathogen is TMV. It can cause tomatoes to become stunted in their growth. Explain how TMV can have this effect. *(2 marks)*

> *TMV interferes with **photosynthesis** ✓. Photosynthesis produces carbohydrate for growth / reduced amount of sugar / starch / energy for growth ✓.*

Protist Diseases: Malaria

Questions on malaria are common and centre around knowledge of parasites, vectors and hosts together with the malarial parasite's life cycle and how the disease can be controlled. There is a link between this section and the section on Sexual and Asexual Reproduction' (page 83), where the parasite's life cycle is described. Take care to learn the terminology and practise applying knowledge of the life cycle, as this next example shows.

Example

Malaria kills many thousands of people every year. The disease is common in areas that have warm temperatures and stagnant water.

a) Explain why malaria is found in these areas. *(2 marks)*

> ✓ ✓ Any two from:
>
> *malaria is transmitted by mosquitos*
>
> *warm temperatures are ideal for mosquitos to thrive*
>
> *stagnant water is an ideal habitat for mosquito eggs to be laid / larvae to survive.*

b) A protist called *Plasmodium* lives in the salivary glands of the female *Anopheles* mosquito.

From the box below, choose a word that describes each organism. *(2 marks)*

parasite	disease	symptom	vector	consumer	host

Mosquito: _____

Plasmodium: _____

> *Mosquito: vector* ✓
> *Plasmodium: parasite* ✓

c) Samit, an African villager, believes that having mosquito nets around the beds of family members and taking antiviral remedies will reduce their risk of catching malaria. Explain why he is only partially correct. *(2 marks)*

> *Nets will deter mosquitoes / prevent bites / prevent transferral of plasmodium* ✓ *but antivirals are ineffective as plasmodium is a protist / not a virus* ✓.

Non-specific Defences

The body has two main systems of defence against infection: non-specific and specific. As the names suggest, non-specific mechanisms generally protect the body and are adapted to prevent pathogens entering living cells. Specific mechanisms are able to target very defined groups of pathogens and are connected with antibodies and their products.

a) Below are the names of some defence mechanisms that the body uses. Match each defence mechanism with the correct function. The first one has been done for you.

(3 marks)

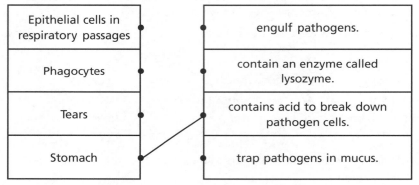

✓ for each correct line up to 3 marks. Subtract 1 mark for any additional lines.

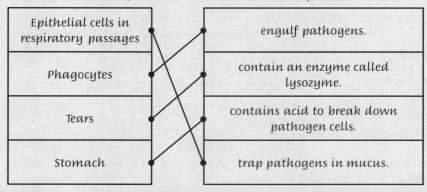

The main point to note here is not to add more lines than the three available. You will lose marks otherwise.

b) Epithelial cells in the trachea trap pathogens in mucus. But this does not get rid of them. Infected mucus can become a problem, causing bronchitis or even pneumonia. How do epithelial cells remove the trapped pathogens? *(2 marks)*

> Beating **cilia** propel mucus and pathogens back up the trachea ✓, where they are then passed into the oesophagus / are swallowed ✓.

Specific Defences and Vaccination

Make sure you understand the difference between the terms **antibody**, **antigen** and **antibiotic**. Here is a handy reference.

- Antibody – specific protein produced in response to a specific antigen on a pathogen. They are part of the body's defence system.

- Antigen – protein on the outside of a foreign cell / particle that can be recognised by antibodies.

- Antibiotic – medicine / drug produced to combat bacterial infections.

Example

The diagram shows a white blood cell producing small proteins as part of the body's immune system.

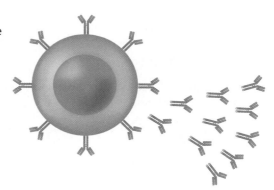

a) What is the name of these proteins? *(1 mark)*

> Antibodies ✓

b) These proteins will eventually lock on to specific invading pathogens. Describe what happens next to disable the pathogens. *(1 mark)*

> Pathogens are clumped together to prevent their further reproduction, to make them easier for phagocytes to digest ✓.

Example

The graph shows the antibody levels in a person after he contracted the flu. The flu pathogen first entered his body two days before point X. There was then a second invasion at point Y.

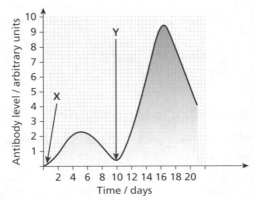

a) Name **one** transmission method by which Dominic could have caught the flu virus. *(1 mark)*

> ✓ Any one from:
> droplet infection / coughing / sneezing / water droplets in breath / aerosol.

b) After how many days did the antibodies reach their maximum level? *(1 mark)*

> Answer in the range 16–17 days ✓.

c) What is the difference in antibody level between point Y and this maximum? Show your working. *(2 marks)*

> 9 arbitrary units ✓ (9.5–0.5) ✓

d) Explain, using your knowledge of memory cells, the difference between these two levels. *(2 marks)*

> Memory cells recognise future invasion of the pathogen ✓; they can produce the necessary antibodies much quicker, and at higher levels, if the same pathogen is detected again ✓ (i.e. when the secondary response occurs).

Vital extra marks are often lost by students because they do not grasp the idea of the body's response to vaccines and forget to mention **memory cells**. Remember, once antibodies have been produced against a particular antigen, the ability to respond quickly to that antigen on the cell surface membrane of a pathogen is built in to the body through memory cell formation. The massive response to a further infection ensures that the body does not suffer any symptoms. This whole process is triggered by vaccination, which is an artificial form of **active immunity**.

Example

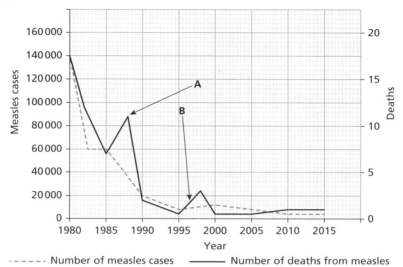

- - - - - Number of measles cases ——— Number of deaths from measles

The graph above shows some data for a country's population between the years 1980 and 2015. Point **A** represents a time when a new vaccine, MMR, was introduced and most new babies received it. MMR stands for measles, mumps and rubella. Point **B** is when large numbers of people chose not to have their babies vaccinated using the MMR vaccine. Some did not have the vaccination at all, while others opted to have their children vaccinated against mumps, measles and rubella separately.

a) Describe the trend relating to the number of measles cases between 1980 and 1995.

(2 marks)

> Measles cases fell overall between these dates ✓. There was a significant rise between 1985 and 1988 ✓.

b) Jan says that the introduction of the MMR vaccine accounts for the fall in cases of measles between 1980 and 1995. Explain the arguments for and against this conclusion.

(3 marks)

> Cases fell significantly when the vaccine was introduced ✓, but cases had also fallen significantly between 1980 and 1985 / before the vaccine was introduced ✓ so other factors could have caused the drop in cases ✓.

c) In 1995 a paper was published that seemed to draw a link between the MMR vaccine and developing autism – a condition characterised by difficulty in communicating and forming relationships with other people. Evaluate the evidence in the graph that the publication of this paper caused an increase in deaths from measles. *(2 marks)*

> *Deaths / measles cases rose after this date* ✓; *this could be coincidence / no causal connection – simply correlation* ✓.

d) Within a year the findings of this paper were found to be suspect and there was a large increase in the uptake of the MMR vaccine again. Radesh says that parents who had their children vaccinated separately left them open to contracting measles. Do you agree? Give reasons for your answer. *(2 marks)*

> No (no marks)
>
> *There is no evidence from the graph that the rise in cases was due to the separate vaccine* ✓.
>
> *Rise in cases could be due to the parents who didn't have their children vaccinated at all / rise could be a coincidence* ✓.

e) Explain why declines in the uptake of a vaccine can be dangerous to the health of a population. *(2 marks)*

> *Loss of herd immunity* ✓.
>
> *More infected individuals are free to pass on pathogens to others* ✓.

f) What is contained within a vaccine? *(1 mark)*

> *A dead / weakened / heat-treated form of the pathogen* ✓.

Antibiotics and Painkillers

The topic of antibiotics reappears in topic 6 – Inheritance, Variation and Evolution. This is because bacteria are becoming increasingly resistant to antibiotics through the process of natural selection. So don't be surprised if this aspect is touched upon in paper 1 too.

The key thing to remember about antibiotics is that they do not work against viruses. Therefore, they should not be prescribed automatically by doctors as this increases the chance of resistance occurring (please note – resistance, not 'immunity' – bacteria do not have an immune system).

Another key point is that current health advice states that the full course of antibiotics should be taken if you are suffering from a bacterial infection.

Example

Isoniazid was a drug developed in 1952 to treat tuberculosis (TB). Today one in seven new cases of TB is resistant to Isoniazid.

a) Explain as fully as you can **one** way in which this resistance could have arisen. *(4 marks)*

> Within the population of TB bacteria there may be a few organisms with natural resistance to isoniazid ✓. This could be a result of a mutation ✓. When a patient is treated with isoniazid, all the sensitive bacteria are killed ✓. This allows resistant bacteria to quickly grow and multiply ✓.

b) Nowadays it is common practice to treat patients with TB using two different antibiotics simultaneously. Explain how this can help reduce antibiotic-resistant strains emerging. *(2 marks)*

> If a bacterium develops resistance to one of the antibiotics ✓, it will still be killed by the second antibiotic ✓.

c) Doctors are concerned about the increase in the number of MRSA (methicillin-resistant *Staphylococcus aureus*) infections they are seeing. What can doctors do to reduce the likelihood of resistant strains emerging? *(2 marks)*

> Not overprescribe antibiotics ✓. Not use antibiotics to treat minor infections ✓.

Painkillers are used to treat the symptoms of disease but do not kill pathogens. In medical language, painkillers are called analgesics. Common painkillers are paracetamol, ibuprofen, aspirin and codeine. These can be obtained over the counter or on prescription. More potent painkillers exist such as morphine. These are used for patients who suffer extreme pain or who need end-of-life care. Questions on painkillers are rare but can form part of more general questions on drugs or clinical trials of drugs.

Discovery and Development of Drugs

Questions on clinical trials are common, and an example is given in this section, but the 'discovery' aspect is quite new. You are likely to be tested on factual recall regarding the discovery of drugs such as penicillin and the process of scientific experimentation and research.

Example

This is amazing. I left some cultures of bacteria to incubate on agar plates while I was on holiday. Upon my return, I found that a couple of the plates had become contaminated with mould colonies. Around the colonies there seemed to be an exclusion zone where no bacteria could grow. I think I may be on to something here.

Alexander Fleming

a) What further work would Fleming need to have done to prove that his discovery was not due to chance or some factor he had not been aware of at the time? *(2 marks)*

> ✓ ✓ Any two from:
>
> *repeat the experiment / conditions in his agar plates*
>
> *record data / observations / measurements many times*
>
> *set up suitable controls*
>
> *publish work for others to check / verify*
>
> *extend study using other strains of bacteria / mould.*

b) In 1928, Alexander Fleming discovered the mould that made the drug which came to be known as penicillin. However, it wasn't until 1945 that mass production and distribution of the drug occurred.

Explain why there was such a long time period between discovery and production.
(3 marks)

> ✓ ✓ ✓ Any three from:
>
> *drug / active ingredient needed to be extracted from the mould / clinical trials needed to be carried out:*
>
> *to see if the drug was safe*
>
> *to see if the drug was effective*
>
> *to determine correct dosage.*

Example

A pharmaceutical company is carrying out a clinical trial on a new drug called alketronol. They are testing it to see whether it produces significant adverse (harmful) events in a sample of 226 patients.

a) Apart from checking for adverse events, write down **two** other reasons why a company carries out clinical trials. *(2 marks)*

> ✓ ✓ Any two from:
>
> to ensure that the drug is actually effective / more effective than a placebo
>
> to work out the most effective dose / method of application
>
> to comply with legislation.

b) The kind of trial carried out is a double blind trial.
What does this term mean? *(2 marks)*

> Double blind trials involve volunteers who are randomly allocated to groups – neither they nor the doctors / scientists know if they have been given the new drug or a placebo ✓.
>
> This eliminates all bias from the test ✓.

c) Data from the trial is shown in the table below.

Adverse event	Number of patients	
	Taking Alketronol	Taking placebo
Pain	4	3
Cardiovascular	21	17
Dyspepsia	7	6
Rash	10	1

i) Calculate the percentage of patients in the trial who suffered a cardiovascular event while taking alketronol. Show your working. *(2 marks)*

> Total patients = 226
>
> $\frac{21}{226} \times 100$ ✓ = 9.29% ✓

ii) A scientist is worried that alketronol may trigger heart attacks. Is there evidence in the data to support this view?
Explain your answer. *(2 marks)*

> *Yes, there are more patients who took the drug and had a cardiovascular event than those who took the placebo. However, there is not a large difference between the groups* ✓.
>
> *No, the cardiovascular events could include other conditions apart from heart attacks* ✓.

iii) Which other adverse event shown in the table might cause concern?
Give a reason for your decision. *(2 marks)*

> *Rash* ✓.
>
> *The difference in numbers of patients who got a rash between the alketronol and placebo groups is quite large* ✓.

Plant Defence Responses

This section is largely about adaptations, so you will need to keep at the forefront of your mind the importance of linking features to advantages, as the example on the next page shows.

Example

The pictures below show two strategies used by plants to deter pests and herbivores.

Foxglove: contains a chemical that lowers heart rate.

Passion flower: possesses structures that look like butterfly eggs.

Explain how effective each of these plant's adaptations are in protecting them against being eaten by **different types of herbivore**. *(3 marks)*

> ✓ ✓ ✓ Any three of the below, with a maximum of 2 marks for one plant.
>
> **Foxglove**
>
> when eaten, the heart is affected, leading to heart attack or lack of oxygen to brain further eating by herbivore's mates or herd deterred as a result
>
> only effective against vertebrates / animals with a heart / ineffective against invertebrates.
>
> **Passion flower**
>
> egg-like structures deter butterflies from laying eggs / butterflies do not want to lay eggs on a flower already with eggs / avoiding competition for the offspring
>
> no caterpillars will hatch
>
> only effective against insects / arthropods
>
> this is an example of **mimicry**.

Notice that to gain full marks in this question you need to comment on **both** examples. A further clue to gaining extra marks is where the question asks you to address different types of herbivore.

 For more on the topics covered in this chapter, see pages 40–45 of the *Collins AQA GCSE Combined Science Revision Guide*.

4

Bioenergetics

Photosynthesis

Photosynthesis is described as an endothermic reaction, meaning that it requires an energy input, in this case from sunlight. The energy is transferred to chloroplasts in the leaves. At the heart of understanding this topic is the photosynthesis equation, and hardly an exam goes by where this basic chemistry is not tested. So, it is well worth your while learning both word and symbol equations 'off by heart'.

Example

In the 17th century a Flemish scientist called Van Helmont carried out some experiments involving weighing the mass of a willow tree over five years. He found that the mass of the tree increased by 30 times and yet the soil mass remained constant.

a) If the tree had a starting mass of 120 g, calculate the finishing mass. Show your working. Show your answer in kilograms.   *(2 marks)*

> 3.6 kg ✓ ✓
> 2 marks for correct answer, but if incorrect, 30 × 0.120 kg would gain 1 mark.

b) Van Helmont concluded that the willow's mass was entirely due to water intake. Explain why Van Helmont was only partially correct.   *(1 mark)*

> *Carbon dioxide is also incorporated from the atmosphere* ✓.

c) What, apart from water, might the willow have absorbed from the soil?   *(1 mark)*

> *Mineral ions / nutrients* ✓

d) Explain why this mass was hardly detectable.   *(1 mark)*

> *Minerals are absorbed in small amounts* ✓.

> This is another example of a question incorporating aspects of scientific discovery and the scientific process. As you practise more of these you will see similar principles appearing time and again; such as a calculation, interpretation of observations and linking your biological knowledge to the experimental findings.

Example

Glucose can be used by plants for energy or to build up bigger molecules. The diagram shows a starch molecule. The part labelled **A** is a glucose molecule.

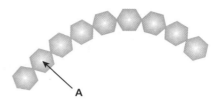

A

a) In which organs of the plant would most of this starch be manufactured? *(1 mark)*

Leaves ✓

b) The plant can synthesise other molecules from the glucose it manufactures, such as cellulose and protein. State **one** use for cellulose and **one** use for protein. *(2 marks)*

Cellulose: cell walls for support ✓.

Protein: growth / cell membranes / enzymes ✓.

c) Write the word equation for photosynthesis. *(2 marks)*

carbon dioxide + water ⟶ glucose + oxygen ✓ For reactants ✓ For products

The Rate of Photosynthesis

One of your required practicals is to investigate the effect of light intensity on the rate of photosynthesis using an aquatic organism such as pondweed. The following question is based on this practical.

Example

Some students were asked to carry out a similar practical, only this time they were asked to investigate the effect of temperature on photosynthesis rate in pondweed. They set up the equipment shown and changed the temperature using ice and hot water. They counted the number of bubbles given off every minute at different temperatures.

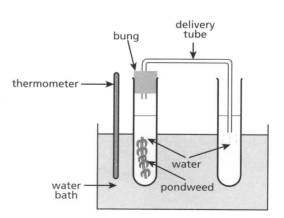

Bioenergetics

a) Why did the students use a water bath? *(1 mark)*

> ✓ Any one from:
>
> to keep the water temperature constant
>
> to vary the temperature for each trial.

b) What should the students have done to make the investigation fair? *(1 mark)*

> Keep all variables (except the one being investigated) the same, e.g. same amount of pondweed for each experiment, same amount of water for each experiment, same light intensity ✓.

c) How could the students make sure their results were reproducible? *(1 mark)*

> Repeat their investigation ✓.

d) Which gas was given off? *(1 mark)*

> Oxygen ✓

e) **HT** The class had to pack up early and didn't finish gathering their data, but they were able to plot a graph (below) from the results they did obtain.

Continue the line on the graph to show the trend you would expect for temperatures above 30°C. *(1 mark)*

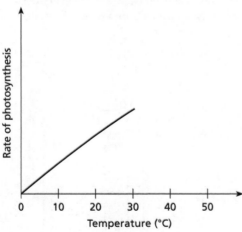

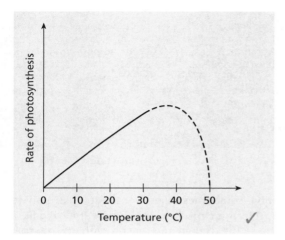

f) **HT** Explain why rate of photosynthesis varies with temperature over the range 0°C to 50°C. *(2 marks)*

As the temperature increases, the rate of photosynthesis increases due to more rapid molecular movement and therefore more successful collisions between molecules ✓. By 37–38°C the rate peaks and beyond this point the enzymes controlling photosynthesis become denatured and the reaction stops ✓.

g) **HT** A market gardener puts a wood-burning stove in his greenhouse to increase the yield of his lettuces.

i) Explain how this will increase the yield. *(2 marks)*

✓ ✓ Any two from:

the increased temperature from the stove will increase the photosynthesis rate

increase in carbon dioxide concentration will have the same effect

increased photosynthesis means increased starch / glucose / carbohydrate production / yield.

ii) Suggest **one** other measure he could take to increase his yield. *(1 mark)*

✓ Any one from:

increase light regime, e.g. artificial lighting switched on at night time

increase light intensity / brighter lights.

Bioenergetics

Higher Tier questions relating to rates of photosynthesis often touch on the idea of **limiting factors**. These factors are temperature, light intensity and carbon dioxide concentration. In general terms, when any one of these variables is increased, the rate of photosynthesis goes up. But this does not continue indefinitely. A maximum rate will be reached and, in the case of temperature, the rate will then fall off due to enzymes becoming denatured. Once the maximum rate is reached, another variable is acting as a limiting factor. In other words, if that factor was increased, the rate would continue to rise again.

Questions of this type are also linked to market gardening where each of these factors can be adjusted to maximise yields.

Respiration

Respiration is not the same as breathing (sometimes called 'ventilation'). Breathing delivers air with high concentrations of oxygen into the lungs where the oxygen is absorbed into the blood. Cells then use the oxygen and react it with glucose to release energy. Beware of describing energy as being created or made. It is better to think of it as changing from one form to another, i.e. from chemical energy to movement, heat or electrical (nervous conduction).

Example

a) Which substance is a product of anaerobic respiration in humans?
   Tick **one** box.   *(1 mark)*

   Carbon dioxide ☐

   Ethanol ☐

   Glucose ☐

   Lactic acid ☐

   > Lactic acid ✓

b) Which substance is a product of aerobic respiration in plants?
   Tick **one** box.   *(1 mark)*

   Carbon dioxide ☐

   Ethanol ☐

   Glucose ☐

   Lactic acid ☐

   > Carbon dioxide ✓

c) Niamh is training for a marathon. Every few days she runs a long distance. This builds up the number of mitochondria in her muscle cells.

What is the advantage for Niamh of having extra mitochondria in her muscle cells? Tick **one** box. *(1 mark)*

Her muscles become stronger. ☐

Her muscles can contract faster. ☐

Her muscles can release more energy. ☐

Her muscles can repair faster after injury. ☐

Her muscles can release more energy. ✓

d) Bob has been running hard and has an oxygen debt. Describe what causes an oxygen debt after a session of vigorous exercise and how Bob can recover from its effects. *(4 marks)*

Cause: build-up of lactic acid (in muscles) ✓ due to anaerobic respiration ✓.
Recovery: heavy breathing / panting ✓, over a period of time ✓.

e) Tariq is competing in a 10-mile running race. His heart rate and breathing rate increase. Describe how this helps his muscles during the race. *(3 marks)*

✓ ✓ ✓ Any three from:

more blood to the muscles

more oxygen supplied to muscles

increased energy transfer in muscles / avoids or reduces anaerobic respiration

more lactic acid is removed from muscles.

Metabolism

Metabolism refers to the sum of all chemical reactions in the body. This includes exothermic (or catabolic) reactions, which release energy, and endothermic (anabolic) reactions, which take energy in. Respiration is the main catabolic reaction in living systems.

HT Example

a) Isaac is running a marathon. Write a balanced symbol equation for the main type of respiration that will be occurring in his muscles. *(2 marks)*

$C_6H_{12}O_6 + 6O_2$ ✓ $\longrightarrow$ $6CO_2 + 6H_2O$ + energy released ✓

Bioenergetics

b) Isaac's metabolic rate is monitored as part of his training schedule. He is rigged up to a metabolic rate meter. This measures the volumes of gas that he breathes in and out.

The difference in these volumes represents oxygen consumption. This can be used in a calculation to show metabolic rate.

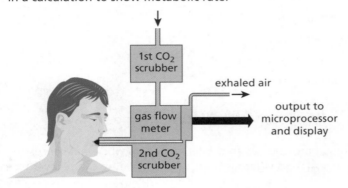

Here are measurements taken from the meter over a period of 1 hour.

	5 minutes of jogging	5 minutes of rest	5 minutes of sprinting	5 minutes sprinting on an incline
Mean metabolic rate per ml oxygen used per kg per min	35	20	45	60

i) The units of metabolic rate are expressed in the table as 'per kg'.

Why is this adjustment made? *(2 marks)*

✓ ✓ Any two from:

larger athletes will use more oxygen due to their higher muscle mass

the adjustment allows rates to be fairly / accurately compared

different athletes may have different masses.

ii) Using the table, explain the difference in readings for jogging and sprinting.

(2 marks)

Sprinting has a greater energy demand ✓, so more oxygen is needed ✓.

iii) Isaac does quite a lot of exercise. His friend Boris does not. How might Boris's readings compare with Isaac's? Give a reason for your answer. *(2 marks)*

Boris's metabolic rate would be lower ✓ because his lungs, heart and muscles are less efficient at transporting / using oxygen ✓.

In this Higher Tier example, the main thing to focus on is interpreting the information in the table. Key to this is understanding the units used for measuring metabolic rate. Oxygen consumption has been chosen because it is the most easily measurable of the reactants in respiration. You may know from chemistry that when assessing reaction rates you can either measure loss of reactants or gain in products.

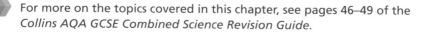

For more on the topics covered in this chapter, see pages 46–49 of the *Collins AQA GCSE Combined Science Revision Guide*.

5

Homeostasis and Response

Homeostasis and Negative Feedback

Control systems in the human body share a lot of comparisons with control systems in chemistry and physics, and you will come across these ideas if you ever study thermostats or biochemical systems. They share three components: receptors / sensors, coordination centres / comparators and **effectors**.

The terms 'homeostasis' and 'negative feedback' are often confused. Homeostasis simply refers to the way in which the body tries to keep certain factors constant, e.g. temperature, blood sugar levels. Negative feedback is a Higher Tier concept and is the mechanism by which the body's internal environment is achieved. The following questions explore these two ideas.

Example

Here is some information about how conditions are kept stable in the human body.

Write down the missing words. Choose **three** words from the box.   *(3 marks)*

effectors	spine	receptors	homeostasis	hormones	glands

Certain factors have to be kept constant in the body. This is achieved by a process

called _____. In order for this to happen, the central nervous system (CNS)

needs to receive information from the environment. This is accomplished through

_____ such as the eye or ear. Once the information has been relayed, the

CNS brings about appropriate changes through _____.

homeostasis ✓; receptors ✓; effectors ✓

Example

Match the structures below, with the function they perform.   *(3 marks)*

Pancreas	Skin receptor	Pituitary gland	Retina

releases ADH	detects pressure	detects light	produces insulin

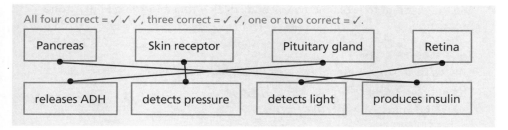

All four correct = ✓ ✓ ✓, three correct = ✓ ✓, one or two correct = ✓.

| Pancreas | Skin receptor | Pituitary gland | Retina |

| releases ADH | detects pressure | detects light | produces insulin |

Example

The diagram below shows how the hormone thyroxine is regulated in the body.

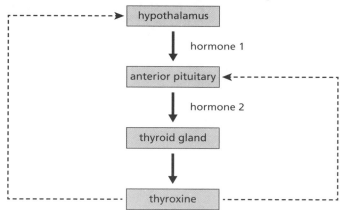

Thyroxine stimulates the basal metabolic rate. It plays an important role in growth and development.

a) What name is given to this process, where a system resists a change from a norm (set point) level? *(1 mark)*

> *Negative feedback* ✓

b) The **pituitary** is often referred to as the 'master gland' of the body because it stimulates other glands to release chemical messengers as well as producing many hormones of its own. Name **one** hormone released by the pituitary. *(1 mark)*

> ✓ Accept one from:
>
> LH
>
> FSH
>
> ADH
>
> These are hormones required by the specification but others might include ACTH, TSH.

c) Use the diagram to explain how the body deals with an increase in thyroxine production. *(3 marks)*

✓ ✓ ✓ Any three from:

more thyroxine in blood detected by hypothalamus

less hormone 1 produced

less hormone 2 produced

negative feedback on hypothalamus and pituitary

fall in thyroxine level.

d) In a condition called hypothyroidism, not enough thyroxine is produced. Symptoms in adults include fatigue, low heart rate and weight gain. Use the information above to explain why these symptoms occur. *(3 marks)*

✓ ✓ ✓ Any three from:

lower metabolic rate lowers heart rate

therefore, less oxygen / nutrients reaching brain

fatigue caused by lack of oxygen / nutrients

lower metabolic rate means less breakdown of carbohydrate

more carbohydrate converted to fat.

Neurones and Reflex Arcs

Questions about this section of the specification often combine features of nerve cells together with how they operate in terms of the reflex arc. If you learn the general sequence below then you can apply it to any number of scenarios. Popular ones include the knee-jerk reflex, pupil reflex and the pain reflex.

Sense organ	Sensory neurone	Synapse	Relay neurone	Synapse	Motor neurone	Muscle
Receptors detect a change either inside or outside the body. This change is a stimulus.	Conducts the impulse from the sense organ towards the CNS.	The gap between the sensory and relay neurones.	Passes the impulse on to a motor neurone.	The gap between the relay neurone and the motor neurone.	Passes the impulse on to the muscle (or gland).	The muscle will respond by contracting, which results in a movement. Scientists call muscles (and glands) effectors.

Example

Merrick has been cooking and has left a hotplate on the cooker on by mistake. He puts his hand down on the hotplate and immediately lifts it off. His response is controlled by a reflex action.

a) What is the stimulus in this reflex? *(1 mark)*

> Heat (from the hotplate) ✓

b) What is the effector? *(1 mark)*

> (Arm) muscle ✓

c) Give **two** characteristics of a reflex action that makes them distinct from other types of response in the body. *(2 marks)*

> Rapid ✓
>
> Unconscious / do not require higher thinking or processing of information ✓.

d) Once Merrick puts his hand on the hotplate, a nervous impulse is passed from his skin receptor along a sensory neurone. Describe the remainder of the pathway that the impulse travels to its completion at the effector. *(4 marks)*

> ✓ ✓ ✓ ✓ Any four from:
>
> sensory neurone carries impulse to relay / intermediate neurone
> in CNS / spine
> impulse passed on to motor neurone
> motor neurone carries impulse to effector / muscle
> reference to synapse / nerve junction.

Notice that this is a description, so you gain marks for simply outlining the route that the nervous impulse takes, using correct terminology. There is no need to explain **how** this happens. This specification does not require you to understand transmission of the impulse at the synapse level.

Reaction Times

One of your required practicals is to plan and carry out an investigation into the effect of a factor on human reaction times. Here is a question to help you see how you might need to draw on this information in an exam.

Example

The reaction times of six people were measured.

They put on headphones and were asked to push a button when they heard a sound. The button was connected to a timer.

Five trials of the experiment were carried out. The table below shows the results.

Person	Gender	Reaction time in seconds				
		1	2	3	4	5
A	Male	0.26	0.25	0.27	0.25	0.27
B	Female	0.25	0.25	0.26	0.22	0.24
C	Male	0.31	1.43	0.32	0.29	0.32
D	Female	0.22	0.23	0.25	0.22	0.23
E	Male	0.27	0.31	0.30	0.28	0.26
F	Female	0.23	0.19	0.21	0.21	0.22

a) Describe a pattern in these results. (1 mark)

Female reaction time better than male / females have faster reaction times than males ✓.

b) Calculate the mean reaction time for person C, ignoring any outliers. Show your working. (2 marks)

$\frac{0.31 + 0.32 + 0.29 + 0.32}{4}$ ✓

Answer is 0.31 ✓.

You are told to ignore any outliers. These values are so far outside the main trend of data as to be considered anomalous and most likely due to error. In this case the number 1.43 is vastly different to the other values.

c) When person D was concentrating on the test, someone touched her arm and she jumped. Her response was a reflex action. What are the **two** main features of a reflex action? *(2 marks)*

> Rapid / fast ✓
>
> Automatic / done without thinking ✓

The Endocrine System

As well as the nervous system, the endocrine system makes up the other method of control available to the human body. Key points include:

- information / signalling is achieved through chemical messengers called hormones rather than nervous impulses
- response times are slower as these hormones have to diffuse to target organs through the bloodstream, which requires time
- hormones often have a more general effect on many cells, tissues and organs
- the response is more drawn out.

Example

Complete the missing information in the table, which is about different endocrine glands in the body. *(4 marks)*

Gland	Hormones produced
Pituitary gland	and
Pancreas	Insulin and glucagon
	Thyroxine
	Adrenaline
Ovary	and
Testes	Testosterone

1 mark for each correct line.

Gland	Hormones produced
Pituitary gland	**TSH, ADH, LH** and **FSH** ✓ Accept any two.
Pancreas	Insulin and glucagon
Thyroid gland ✓	Thyroxine
Adrenal gland ✓	Adrenaline
Ovary	**Oestrogen** and **progesterone** ✓
Testes	Testosterone

Control of Blood Sugar and Diabetes

This area of control illustrates how homeostasis and negative feedback operate in a specific example. Higher Tier knowledge is required to explain the role of the additional hormone, glucagon, in negative feedback.

Example

A new nanotechnology device has been developed for people with diabetes – it can detect levels of glucose in the blood and communicate this information to a hormone implant elsewhere in the body. The implant releases a precise quantity of hormone into the bloodstream when required.

a) Explain how this device could help a person with type 1 diabetes who has just eaten a meal. *(2 marks)*

> ✓ ✓ Any two from:
> after a meal, a rise in glucose levels will be detected by device
> which will cause hormone implant to release insulin
> insulin released to bring blood glucose level down.

b) Explain why a person with type 2 diabetes might not have as much use for this technology. *(2 marks)*

> People with type 2 diabetes can often control their sugar level by adjusting their diet ✓; body's cells often no longer respond to insulin ✓.

> In questions about diabetes, make sure you refer to blood glucose as glucose is found all over the body, but it is in blood that the control system operates.

HT Example

The word equation below shows how two hormones produced in the pancreas regulate blood glucose in the body.

glucose $\xrightleftharpoons[\text{glucagon}]{\text{insulin}}$ glycogen

Lazlo has just eaten a doughnut before spending an hour playing a game of basketball.

Use the equation to describe and explain how negative feedback operates to ensure that his blood glucose levels remain within acceptable limits over this period of time.

(4 marks)

✓ ✓ Any two from:

glucose levels in blood raised after doughnut

changes detected (in the pancreas)

insulin released from cells in pancreas

stimulate body cells to absorb glucose

also stimulate liver and muscle cells to convert glucose to glycogen.

✓ ✓ Any two from:

blood glucose levels fall during the basketball game

changes detected in pancreas (mark available if not already gained above)

glucagon released from different cells in pancreas

this hormone stimulates liver and muscles to convert glycogen to glucose.

It is important not to confuse hormones with enzymes. Although they are proteins they have a different mode of action. The three 'g' words can also be confusing: glucose – the sugar, glycogen – the storage carbohydrate, and glucagon – the hormone.

Be aware that some questions about blood glucose control may ask you to interpret graphs showing these changes in someone's blood over time. Comparing diabetics with non-diabetics is also a possibility.

Reproductive Hormones

Reproductive hormones are another complex area as the interplay of female hormones involves four chemical messengers: FSH, oestrogen, progesterone, and LH.

Example

The graph shows the thickness of the uterus during the menstrual cycle. Use the graph and your scientific knowledge to describe what happens in the woman's ovaries and uterus between days 5 and 28. *(3 marks)*

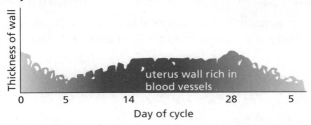

Days 5–14: uterus wall is being repaired ✓, egg released at approximately 14 days from ovary ✓, days 14–28: uterus lining maintained ✓.

HT Example

The diagram shows how the human process of ovulation is controlled.

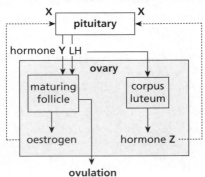

a) X represents an effect that two hormones have on the pituitary gland. Write down the name of this effect. *(1 mark)*

Negative feedback ✓

b) Name hormone Y. *(1 mark)*

FSH ✓

c) Name hormone Z. *(1 mark)*

> *Progesterone* ✓

d) Describe the effect on hormone Z if the egg is fertilised. *(2 marks)*

> *Continues to be produced* ✓ *in large quantities / at high levels* ✓*.*

> The best way to become confident in explaining changes in these processes is to use a generalised diagram and imagine various scenarios that occur during the woman's monthly cycle. Try following the arrows on the hormone flow diagram above and say aloud what would happen when one hormone level is raised. Repetition of this exercise will embed the information.

Contraception and Infertility Treatment

Tackling questions on contraception and infertility means that you will need a thorough understanding of the functions of female (and sometimes male) hormones. You will often find ethical considerations as part-questions.

Example

Tim and Margaret are finding it hard to conceive a child. They visit a fertility clinic and meet some other couples. The table shows some information about the problem that each couple has.

Couple	Problem causing infertility	Percentage of infertile couples with this problem	Percentage success rate of treatment
Tim and Margaret	Blocked fallopian tubes	13	20
Rohit and Saleema	Irregular ovulation	16	75
Leroy and Jane	No ovulation	7	95
Gary and Charlotte	Low sperm production	15	10
Ian and Kaye	No sperm production	21	10
Stuart and Mai	Unknown cause	28	–

Homeostasis and Response

a) Which couple has the best chance of being successfully treated? *(1 mark)*

> Leroy and Jane ✓

b) In how many of the six couples is the problem known to be with the female? *(1 mark)*

> Three: Tim and Margaret, Rohit and Saleema, and Leroy and Jane ✓.

c) The treatment of irregular ovulation and no ovulation have the highest success rates.

Explain why treating irregular ovulation would produce more pregnancies in the whole population. *(2 marks)*

> Although irregular ovulation has a lower success rate ✓, it affects over twice as many couples (16 × 75 produces a larger total than 7 × 95) ✓.

d) Leroy and Jane are considering two methods to help them have children. The first is to have an egg donated by another woman. The second is to arrange for another woman to conceive the child using sperm from Leroy, then give birth to it (surrogacy). What are the advantages and disadvantages of each method? *(4 marks)*

> Both methods mean that Jane does not make any genetic contribution ✓.
> Egg donation has a high rate of success but can be expensive ✓.
> Surrogacy might be cheaper but there is a risk that the surrogate mother might develop an attachment to the baby / want to keep it ✓.
> Egg donation requires invasive technique ✓.

Example

Explain how the contraceptive pill works. Name any hormones involved. *(2 marks)*

> ✓ ✓ Any two from:
> the contraceptive pill contains hormones that inhibit FSH production
> e.g. oestrogen / progesterone
> eggs therefore fail to mature
> progesterone causes production of sticky cervical mucus that hinders movement of sperm.

For more on the topics covered in this chapter, see pages 50–55 of the *Collins AQA GCSE Combined Science Revision Guide*

6

Inheritance, Variation and Evolution

Sexual and Asexual Reproduction

Reproduction is one of the seven characteristics of life, and at GCSE you will have studied well beyond the subject of human reproduction and considered the range of reproductive strategies that other organisms adopt. Remember, reproduction drives survival, which in turn drives evolution. So it is no wonder that these ideas are linked together in one topic.

Example

From the box below, choose **three** words to complete the sentences.   *(3 marks)*

zygotes	gametes	diploid	haploid	mitosis	meiosis

Eggs and sperm are known as sex cells or _____. They are described

as _____ because they contain one set of chromosomes. Eggs and

sperm are produced in the ovaries and testes by _____.

> Eggs and sperm are known as sex cells or **gametes**. They are described as **haploid** because they contain one set of chromosomes. Eggs and sperm are produced in the ovaries and testes by **meiosis**.

Example

Which statements about causes of variation are true? Tick **two** boxes.   *(2 marks)*

Meiosis shuffles genes, which makes each gamete unique.   ☐

Gametes fuse randomly.   ☐

Zygotes fuse randomly.   ☐

Mitosis shuffles genes, which makes each gamete the same.   ☐

> Meiosis shuffles genes, which makes each gamete unique ✓.
> Gametes fuse randomly ✓.

DNA and the Human Genome Project

Foundation Tier knowledge of DNA covers basic DNA structure, the nature of DNA as a code in terms of base sequences, historical aspects of DNA's discovery, the Human Genome Project and the Genographic Project. If you learn the foundation ideas about DNA and what its function is, you should be able to apply these to questions that ask you to use your knowledge in novel situations.

Example

Which statements about the Human Genome Project (HGP) are true?
Tick **three** boxes.   *(3 marks)*

The genome of an organism is the entire genetic material present in its adult body cells. ☐

The data produced from the HGP produced a listing of amino acid sequences. ☐

The HGP involved collaboration between US and UK geneticists. ☐

The project allowed genetic abnormalities to be tracked between generations. ☐

The project was controversial as it relied on embryonic stem cells. ☐

The genome of an organism is the entire genetic material present in its adult body cells. ✓

The HGP involved collaboration between US and UK geneticists. ✓

The project allowed genetic abnormalities to be tracked between generations. ✓

Example

Studies of genomes can help scientists work out the evolutionary history of organisms by comparing the similarity of particular DNA sequences that code for a specific protein.

The table shows the percentage DNA coding similarity for protein A in different species.

Species	% DNA coding similarity between species and humans for protein A
Human	100
Chimpanzee	100
Horse	89
Fish	79
Yeast	67
Protist	57

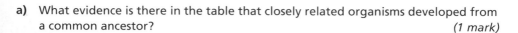

a) What evidence is there in the table that closely related organisms developed from a common ancestor? *(1 mark)*

> *Organisms with very similar features / chimpanzees and humans share equal DNA coding for protein A ✓.*

b) Using only the information from the table, which invertebrate is the most closely related to humans? *(1 mark)*

> *Yeast ✓*

Example

The Human Genome Project has enabled specific genes to be identified that increase the risk of developing cancer in later life. Two of these genes are the *BRCA1* and *BRCA2* mutations. If women are prepared to take a genetic test, how could this information help doctors advise women about breast cancer? *(2 marks)*

> ✓ ✓ Any two from:
> *warn women about the risk of cancer ahead of time*
> *enable early and regular screening*
> *enable early treatment*
> *suggest treatment that is targeted.*

Inheritance: Working out Genetic Crosses

Working out genetic crosses is a little like calculating maths problems. You understand the ideas behind them, grasp the constructs (arrow diagrams / Punnett squares) and then carry out many examples to get the hang of them. Before attempting these problems, make sure you understand the meaning of the following terms: **phenotype**, **genotype**, **recessive**, **dominant**, **heterozygous** and **homozygous**.

Generally, three types of 'cross' appear in questions. These could be called the 'pure breeding' cross, which yields only the dominant phenotype, the 'heterozygous' cross, which gives a 3:1 ratio of phenotypes, and the 'back-cross', which produces a 1:1 ratio of phenotypes. We will explore all three types of cross in the following examples.

Inheritance, Variation and Evolution

Example

Raj is the owner of two dogs and both are about two years old. Both dogs are black in colour and came from the same litter of puppies.

a) A dog's adult body cell contains 78 chromosomes. How many chromosomes would be in a male dog's sperm cells? *(1 mark)*

> 39 ✓

b) The dogs' mother had white fur and the father had black fur. Using what you know about dominant alleles, suggest why there were no white puppies in the litter. *(2 marks)*

> *Black is the dominant gene / allele; white is recessive* ✓. No marks given for references to 'black chromosome' or 'white chromosome'.
>
> *The allele for black fur is passed on / inherited from the father* ✓.

c) **HT** One year later, one of the black puppies mated with a white-haired dog. She had four puppies. Two had black fur and two had white fur.

The letters **B** and **b** represent the alleles for fur colour: **B** for black fur and **b** for white fur.

Draw a fully labelled genetic diagram to explain this. Show which offspring would be black and which would be white. *(3 marks)*

	b	**b**
B	Bb Black	Bb Black
b	bb White	bb White

Correct genotype or gametes for both parents (Bb and bb) ✓

Genotype of offspring correct (Bb and bb) ✓

Correct phenotype of offspring ✓

Or

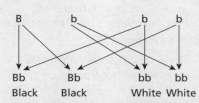

Bb   Bb   bb   bb
Black   Black   White   White

> Part **c)** of this question is an example of a 'back-cross.' For Foundation Tier, you will be provided with the Punnett square for working out a genetic cross. For Higher Tier, you are expected to construct your own. Remember to represent the alleles as single letters and to make the distinction between capital (dominant) and lower case (recessive) letters. Some letters cause particular difficulty, e.g. N and n.

Lastly in this section, let's look at the heterozygous cross. A popular example is the inheritance of eye colour.

HT Example

Explain how parents who both have brown eyes could produce a child who has blue eyes. Use a genetic diagram to help you, and state the probability of a brown-eyed child occurring. *(4 marks)*

If both parents are heterozygous for this gene, e.g. Bb ✓, then there is a 1 in 4 chance of the child having blue eyes ✓.

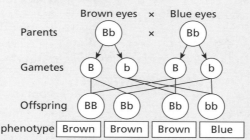

✓ ✓ for correct diagram. Subtract 1 mark for every incorrect row.

The above answer makes use of an 'arrows' diagram. You could equally have used a Punnett square.

Genetic Defects

This section will deal with how to tackle questions involving genetic abnormalities or defects. The principles of using the genetic cross diagrams still apply but the only difference is recognising and interpreting how the defect expresses itself.

The following example highlights the popular examples, **cystic fibrosis** and **polydactyly**, as these are required by the specification.

Inheritance, Variation and Evolution

Example

Fill in the missing words to complete the following sentences.

Choose from the words below. *(4 marks)*

heterozygous	dominant	carriers	two
recessive	one	three	homozygous

Polydactyly is a condition that causes extra fingers or toes. It's caused by a

_____ allele. Only _____ parent needs to have the

disorder for a child to be affected.

> Polydactyly is a condition that causes extra fingers or toes. It's caused by a **dominant** allele. Only **one** parent needs to have the disorder for a child to be affected. ✓ ✓

Cystic fibrosis is caused by a _____ allele. It must be inherited

from both parents. The parents might not have the disorder, but they might be

_____.

> Cystic fibrosis is caused by a **recessive** allele. It must be inherited from both parents. The parents might not have the disorder, but they might be **carriers**. ✓ ✓

Example

A man who knows he is a carrier of cystic fibrosis is thinking of having children with his partner. His partner does not suffer from cystic fibrosis, neither is she a carrier. Suggest what a genetic counsellor might advise. Use genetic diagrams to aid your explanation and state any probabilities of offspring produced. *(6 marks)*

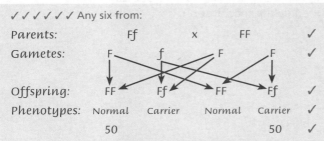

✓ ✓ ✓ ✓ ✓ ✓ Any six from:

Parents: Ff   x   FF   ✓

Gametes: F   f   F   F   ✓

Offspring: FF   Ff   FF   Ff   ✓

Phenotypes: Normal   Carrier   Normal   Carrier   ✓

       50   50   ✓

No children will suffer from cystic fibrosis ✓.

50% chance of a child being a carrier ✓.

Sex Determination

This area of genetics should be straightforward to understand, yet many students lose marks due to incorrectly stating that the male genotype is simply 'Y', whereas it is 'XY'.

Example

What is produced from the fusion of two sex cells? *(1 mark)*

A zygote ✓

Example

Which of the following are the female sex chromosomes, and which are the male sex chromosomes? Label them correctly. *(1 mark)*

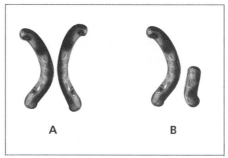

A: female, B: male ✓

Example

John and Salmyra decide to have children. Use a genetic cross diagram to show that the probability of them producing a girl is 50%. *(4 marks)*

Parents: John XY   x   Salmyra XX   ✓
Gametes: X   Y   x   X   X   ✓
Offspring:

	X	Y
X	XX	XY
X	XX	XY

✓

Ratio of male:female   1:1 ✓
(Probability of female child therefore 50%.)

Variation and Evolution

Variation, as far as GCSE biology is concerned, refers to differences in individuals within the same species. From KS3 work you will know that this can either be caused by the genes organisms possess, the influence of their environment, or both. Variation is closely linked to evolution because it drives the process called **natural selection** (the mechanism by which evolution occurs). This next example explores these connections.

Example

Scientists frequently study the distribution of the common snail, *Cepaea*. The snail has a shell that can be brown or yellow, and striped or unstriped. The shell colour and banding influences the visibility of snails to thrushes that prey on them.

A recent study compared the distribution of snails in forest and countryside areas. The results, as a percentage, are shown below.

Area	Striped shell (%)	Unstriped shell (%)
Forest / woodland	13	87
Open countryside / hedgerows	75	25

a) Suggest a reason for this distribution of snails. *(2 marks)*

> More striped snails in countryside / hedgerows ✓ because the shell blends in with the striped nature of the vegetation ✓.

b) Which plain-coloured snail would you expect to find most of in forest / woodland areas? Explain your answer. *(2 marks)*

> Brown ✓. Brown camouflages more easily with the earth / forest-floor ✓.

Snails are part of a larger group of invertebrates called gastropods. It is thought that they evolved from a primitive group of molluscs called lophophores. This group lived underwater and filter fed on smaller marine organisms using a ring of tentacles around their mouths. Snails, on the other hand, are land-dwelling and feed on plant material using a rasping tongue or radula.

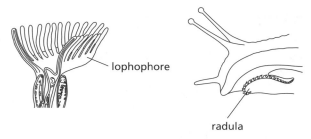

radula

c) Explain, using ideas about natural selection, how a population of sea-dwelling lophophores could give rise to an ancestor of plant-eating gastropods in the oceans. *(4 marks)*

✓ ✓ ✓ ✓ Any four from:

mutation in lophophore population, which lacks tentacles round mouth but has more rasping mouthparts

new individual can exploit new habitats / feed off different foods

change in the environment / habitat means that there is more plant food than other sources of food; new individual's mouthparts are more adapted to feeding on plant material

they outcompete the lophophores

gastropods are more competitive

and survive to pass on mutated genes to next generation

large numbers of new individuals outcompete old population leading to the original species' extinction.

d) Suggest what may happen to the numbers of brown- and yellow-shelled snails if our climate continues to get hotter due to global warming. Give a reason for your answer. *(2 marks)*

✓ ✓ Any two from:

fewer brown-shelled / more yellow-shelled snails

brown-shelled snails overheat more easily

less competition for yellow-shelled snails.

e) Both these shell types are found within the same species. Suggest how this could have occurred. *(2 marks)*

Mutation / change in genes coding for shell colour ✓; different protein (for shell pigmentation) produced ✓.

Natural Selection

Although these ideas are much more complex than this example, a good starting point to understanding Darwin's ideas is to think of evolution as a car, in that it is the vehicle that allows us to get from A to B; or to see species X become organism Y over millions of years. Natural selection could therefore be likened to the engine in the car, in other words it is the mechanism by which the car moves. When explaining how natural selection operates, it is useful to keep in mind the natural selection 'template'. Each factor in this template was originally conceived by Charles Darwin. These factors are:

- variation
- competition
- survival of the fittest
- inheritance
- extinction.

You will gain full marks if you include all five of these factors in your answer.

Example

The picture below shows the limbs of five species of vertebrate. They are all based on the **pentadactyl** limb, which means a five-digited arm, leg, wing or flipper.

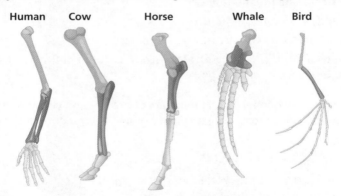

Human   Cow   Horse   Whale   Bird

Scientists believe that the whale may have evolved from a horse-like ancestor that lived in swampy regions millions of years ago. Suggest how whales could have evolved from a horse-like mammal. In your answer, use Darwin's theory of natural selection. *(5 marks)*

✓ ✓ ✓ ✓ ✓ Any five from:

Variation – horse-like ancestor adapted to environment / had different characteristics; named examples of different characteristics, e.g. some horse-like mammals had more flipper-like limbs (as whales have flippers); mutation in genes allowed some individuals to develop these advantageous characteristics.

Competition for limited resources – examples of different types of competition, e.g. obtaining food in increasingly water-logged environment.

Survival of the fittest – named examples of different adaptations, e.g. some horse-like mammals had more flipper-like limbs that allowed them to swim better in water than those who did not.

Inheritance – genes from the flipper-like individuals passed on to the next generation as they survived and the others did not.

Extinction – variety that is least successful does not survive and becomes extinct.

Take care not to word your explanation as if it was Lamarckism. Remember, the genetic change comes first – in this case genes that code for a more flipper-like pentadactyl limb. It is not the case that the horse-like ancestor recognised the need to have flippers and therefore evolved them! It may be possible for you to gain marks by explaining in more detail how speciation occurs, but this is dealt with in the next section.

Selective Breeding

Natural selection is often easier to understand if you first cover selective breeding. It seems a straightforward thing to appreciate that humans have chosen characteristics they want in both plants and animals, then bred them to achieve these characteristics. With natural selection, it is simply that nature or the principle of 'survival of the fittest' provides the driving force rather than human choice.

When describing artificial selection or selective breeding, keep in mind the general template:

- humans select the desired characteristic in parents
- the selected individuals are encouraged to breed together (at the exclusion of other individuals) or cross-pollinate in the case of plants
- the offspring resulting from the mating can then be chosen and the whole process repeated again.

Questions ask you to apply this template and often to mention drawbacks or compare it with the process of genetic engineering (see the next section).

Example

Humans have selectively bred Golden Retrievers so that they can be used as guide dogs for the visually impaired (VI). They have calm temperaments and are easily trained.

a) Describe how a dog-breeder might consistently produce retrievers useful for these training purposes. *(3 marks)*

> ✓ ✓ ✓ Any three from:
>
> *choose adult retrievers with good temperament and ability to follow training*
>
> *arrange for these individuals to breed together*
>
> *allow puppies to grow to adulthood*
>
> *choose dogs that show the desired characteristics*
>
> *breed with other high quality dogs (from a different family)*
>
> *repeat the process many times.*

b) Golden retrievers from the UK are more than twice as likely to get lymphoma as other breeds of dog. Some say this is due to the selective breeding process. Explain why this might occur and discuss whether such breeding should continue in the future. *(3 marks)*

> ✓ Any one from:
>
> *defective alleles more likely to be paired together in close relatives*
>
> *cancer / lymphoma more likely with these genes.*
>
> ✓ Any one from:
>
> *cruel to limit dog's life expectancy in this way*
>
> *less variation in retrievers / reduces gene pool.*
>
> ✓ Any one from:
>
> *dogs with these characteristics greatly increase quality of life for VI people*
>
> *better to allow this type of selective breeding under controlled conditions*
>
> *more expensive to manage mobility for VI people by other means*
>
> *allows companionship for VI people.*

Example

Give **two** examples of characteristics that farmers might want to selectively breed into their crops. *(2 marks)*

> Disease-resistant crops ✓.
>
> Large or unusual flowers ✓.
>
> (Other characteristics not included on the specification: drought-resistance; high-yield cereals.)

Genetic Engineering

Compared with selective breeding, genetic engineering is very precise and characteristics can be produced exactly in the desired organism. Genetic engineering produces genetically modified (GM) organisms – sometimes called transgenic organisms. This has led to controversy regarding the type of organisms and products produced. Questions on the drawbacks, both real and imagined, might appear in questions you have to answer.

Example

Describe how genetic engineering is different from selective breeding. *(2 marks)*

> ✓ ✓ Any two from:
>
> involves genes, not whole organisms
>
> genes transferred from one organism to another
>
> more precise in terms of passing on characteristics
>
> rapid production / process.

Example

Explain the benefit of each of these examples of genetic engineering.

a) Resistance to rose black spot fungus in roses. *(2 marks)*

> Less manually intensive / no need to remove leaves, etc. ✓.
>
> Saves money on costs / economic advantages in terms of profit ✓.

b) Introducing genes into oranges to produce larger fruit. *(1 mark)*

> Greater yield for similar input of resources from the grower ✓.

Inheritance, Variation and Evolution

Example

Some people think that genetically engineering resistance to herbicides in plants could have harmful effects. Give **one** example of a harmful effect. *(1 mark)*

> GM plants may cross-breed with wild plants, resulting in wild plants / weeds that are herbicide-resistant or that outcompete other native species ✓.

Example

a) Describe the process where artificial insulin is produced via genetic engineering.

State any enzymes involved and how the final product is obtained.

The first stage has been completed for you. *(4 marks)*

<u>Human gene for insulin identified.</u>

> Gene removed using **restriction enzyme** ✓.
> Bacterial plasmid 'cut open' using restriction enzyme ✓.
> **Ligase enzyme** used to insert human gene into bacterial plasmid ✓.
> Insulin **purified** from fermenter culture and produced in commercial quantities ✓.

b) The 'cut' gene is inserted into a bacterium. Why are bacteria good host cells for the 'cut' insulin gene? *(2 marks)*

> They reproduce rapidly; can be grown in large vats economically ✓.
> Produce large quantities of insulin in a short time ✓.

c) What remaining step is required before the insulin can be used by patients? *(1 mark)*

> Insulin needs to be purified / separated from bacteria and reactants ✓.

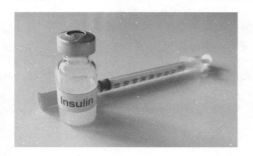

Antibiotic Resistance

This section links with Chapter 3: Infection and Response. This time the emphasis is on how microorganisms such as MRSA could have evolved to become resistant to antibiotics.

Example

Antibiotics are becoming increasingly ineffective against 'superbugs' such as MRSA.

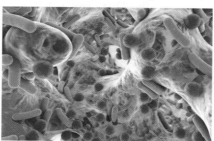

a) Use ideas about natural selection to explain how this has occurred. *(4 marks)*

✓ ✓ ✓ ✓ Any four from:

bacteria mutate – gene for resistance arises randomly

non-resistant bacteria more likely to be killed by antibiotics / are less competitive

antibiotic-resistant bacteria survive and reproduce often

resistant bacteria pass on their genes to the next generation

gene becomes more common in general population

non-resistant bacteria are replaced by newer, resistant strain.

b) Describe how medical practitioners can help reduce this problem of resistance. *(2 marks)*

Do not prescribe antibiotics for viral infections ✓.

Patients should complete the entire course of antibiotics ✓.

Inheritance, Variation and Evolution

Evidence for Evolution

a) This diagram shows the jellyfish species, one of the first multicellular organisms to evolve.

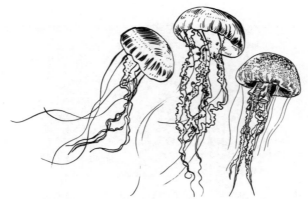

Evidence for jellyfish in the fossil record is rare. Explain why this is. *(2 marks)*

> They are soft-bodies ✓.
> Soft body parts decay and are therefore not preserved ✓.

b) Vertebrates are more easily fossilised. After decay of the soft parts, what stages have to occur for hard parts to be fossilised and then discovered? *(3 marks)*

> Rock sediments compress organism's remains ✓.
> Hard parts replaced by minerals ✓.
> Tectonic movements bring remains to surface ✓.

c) How can fossils be used to identify how one form may have evolved into others? *(1 mark)*

> Compare fossil forms and where they are found in the rock layers (of different ages) ✓.

d) Scientists believe we are about to witness the sixth greatest extinction event in the history of planet Earth. How is this event likely to differ from previous ones in terms of its causes? *(1 mark)*

> Extinction caused by the impact of human-caused climate change ✓.

Classification

Classification simply means the sorting of things into groups, in this case, organisms.

Example

a) In the Linnaean system of classification, organisms are placed within a hierarchy of organisation. The following table shows this hierarchy with some missing levels. Complete the missing levels.   *(2 marks)*

Kingdom
Class
Family
Genus
Species

Phylum ✓
Order ✓

b) Linnaeus also developed the **binomial system** of naming organisms. The domestic cat has the following classification:

Kingdom – Animalia

Class – Mammalia

Family – Felidae

Genus – Felis

Species – Catus

Use this information to suggest the binomial name of the common cat.   *(1 mark)*

Felis catus ✓

Inheritance, Variation and Evolution

Archaeopteryx is an ancient fossilised bird. When first discovered, scientists found it hard to classify.

c) i) Using features shown in the picture, explain why *Archaeopteryx* is difficult to classify. *(2 marks)*

> Possesses features that are found in both reptiles and birds ✓.
>
> Feathers place it with birds but it also has teeth / does not have a beak like reptiles – it is an intermediate form ✓.

ii) Carl Woese developed a new classification system that divided organisms into three main groups or domains. In which group would *Archaeopteryx* be placed? *(1 mark)*

> Eukaryota ✓

iii) Scientists used to use external features to classify organisms. What technological developments have allowed more accurate systems like those of Woese to be developed? Describe how they have aided this development. *(2 marks)*

> Microscopes – to allow observation of internal features. ✓
>
> Biochemical processes – molecules common to different organisms suggest closer relationships. ✓

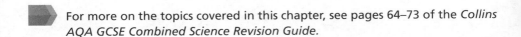

For more on the topics covered in this chapter, see pages 64–73 of the *Collins AQA GCSE Combined Science Revision Guide*.

7 Ecology

Communities – Interdependence

Ecology literally means the study of the 'homes' of organisms, although we think of the places where they live as niches, habitats, ecosystems or even the whole biosphere. Each of these terms needs to be understood in order to appreciate how organisms interact. And this is why 'interdependence' is the starting point for this chapter. It emphasises that all organisms are interconnected through relationships such as competition, predation and shared living space.

Like all the ideas covered in this biology booster guide, ecology has its basic rules and principles, but they are often more difficult to grasp because they operate on such a grand scale. We will try to pick our way through this maze and focus on the key points that examiners look for.

Example

Bird populations are a good indicator of environmental sustainability and they allow scientists to track environmental changes in particular habitats.

Scientists measured the numbers of farmland birds and woodland birds in the UK between 1972 and 2002.

Their results are shown below.

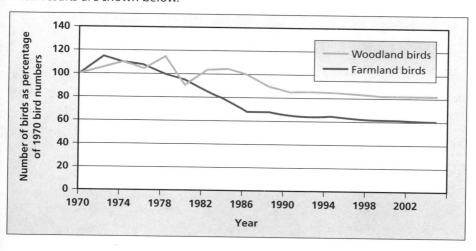

Ecology

a) Describe how numbers of farmland and woodland birds have changed between 1972 and 2002. *(4 marks)*

✓ ✓ ✓ ✓ Any four from:

overall decline in both species

initial rise in farmland birds in 1974

woodland birds show fluctuations / peaks and troughs between 1970 and 1986

rapid decline in farmland birds from 1974 onwards

more gradual decline in both species from 1986.

Notice that you only need to describe these trends. There is no need to explain why they might have happened. That is the subject of part **c)**.

b) Suggest a reason for the overall change in numbers of farmland birds. *(1 mark)*

Farmers have cut down hedgerows and / or trees so the birds have had nowhere to nest and their food source has been reduced ✓.

c) The government wants to reverse these changes by 2020. Suggest **one** thing it could do that would help to achieve this. *(1 mark)*

✓ Any one from:

plant more trees

encourage farmers to plant hedgerows

encourage farmers to leave field edges wild as food for birds

use fewer pesticides.

d) State **two** factors that animals living in the same habitat will compete for. *(2 marks)*

✓ ✓ Any two from:

food

mates

territory.

Example

A survey of insect life was carried out on three nature reserves found in three areas of the UK.

The survey assessed the numbers of three species of insect: two butterflies (species A and B) and a parasitic wasp (species C).

Both species A and B feed on the same plant, a broad-leaved herb found in decreasing amounts on the reserves.

The parasitic wasp feeds on the caterpillars of species A and B by laying its eggs inside the live organism.

The graph below shows the proportions of each species found in the three reserves as a percentage of the total of the three populations. The 'boundary' zone is found between the northerly reserve and the southerly reserve. The average Summer temperatures for each region are also shown.

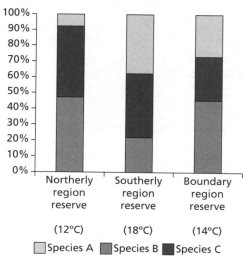

a) Using ideas about competition and predation, compare the proportions of each species for the three regions and suggest reasons for these differences. *(5 marks)*

✓✓ Any two from:

species B more numerous in north and boundary region / species A less numerous in these regions; species A more numerous in the South; species C (wasp) similar proportions in north and south; species C (wasp) less numerous in the boundary region.

✓ Any one from: **Northerly reserve** – species B outcompetes species A; species A preyed upon more by species C (wasp); lower temperatures favour species B.

✓ Any one from: **Southerly reserve** – species A outcompetes species B; species B preyed upon more by species C (wasp); higher temperatures favour species A.

✓ Any one from: **Boundary reserve** – species A is higher in number than in the northerly region as the temperature is not too cold / parasitic wasp is fewer in number; species B similar numbers compared to northerly region as not too cold; species C is fewer in number because the intermediate temperature favours **both** species A and species B.

This question might appear daunting at first, but there are many ways to pick up the marks by simply stating the patterns from the graph. Also, you are given a clue to the reasons in the stem of the question. You don't have to be more detailed about why competition and predation operates as it does in this situation. Please note that in order to gain maximum marks, you must address both the comparisons and the reasons parts of the question.

Be aware that there are many factors that affect the abundance and distribution of organisms – both **biotic** and **abiotic**. Take time to familiarise yourself with each of these categories. They include:

- biotic – availability of food, new predators, new pathogens
- abiotic – light intensity, temperature, moisture levels, soil pH and mineral content, wind intensity and direction, carbon dioxide levels for plants and oxygen levels for aquatic animals.

Adaptations in Animals

When describing adaptations it is essential to state the biological feature of the organism and **why** that feature makes it well adapted. The following questions deal with both plant and animal examples.

Example

The frilled lizard lives in Australia and feeds off cicadas, beetles, termites and mice. Its habitat exposes it to high daytime temperatures and it can often be seen basking in the morning. The frill round its neck can be quickly erected to give a startling display that can distract other animals. It can climb trees expertly and runs quickly on its hind legs.

Explain how the lizard's features make it well adapted to its environment.   *(3 marks)*

Frill can be erected to startle / scare predators / give it time to escape ✓.

Running on hind legs allows it to escape predators / climbing ability allows it to escape predation ✓.

Basking in the sunlight helps it to warm up for daytime activity / catching prey ✓.

Example

Scientists have discovered shrimps and giant worms that live clustered around hot vents on the ocean floor. These organisms can survive temperatures of up to 110°C. What name is given to such organisms? Underline the correct answer.   *(1 mark)*

mesophiles   gravophiles   extremophiles

extremophiles ✓

Plants

Example

Dandelions, docks and thistles are all weeds and are well adapted to compete with other plants. The drawing below shows some features of a thistle.

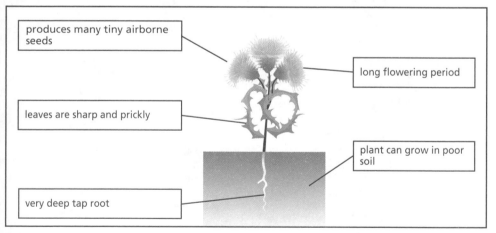

produces many tiny airborne seeds

long flowering period

leaves are sharp and prickly

plant can grow in poor soil

very deep tap root

Choose any **three** of the above features. For each feature explain how it helps the thistle to compete and survive.   *(3 marks)*

✓ ✓ ✓ Any three from:

many tiny airborne seeds – seeds can be carried a long way and even if many die, plenty will grow into adult plants

sharp, prickly leaves – this will deter animals from eating the thistle

very deep tap root – allows plant to get more water / makes it difficult to pull up

long flowering period – plenty of opportunity for insects to pollinate plants

plant can grow in poor soil – it can grow where other plants might not be able to.

Food Chains and Webs

At KS3 you will have learned about food chains, webs and even pyramids of numbers. At GCSE you are expected to apply these ideas further to understanding how energy flows through food chains, and how this has implications for farming and other types of food production.

Example

The food web below exists in a freshwater pond habitat.

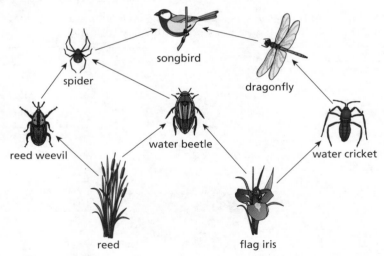

a) Write out **two** food chains; one involving three organisms, and the other involving four organisms. Each chain must include a producer, one primary consumer and one secondary consumer. The second chain must include an apex predator. *(2 marks)*

_____ ⟶ _____ ⟶ _____

_____ ⟶ _____ ⟶ _____ ⟶ _____

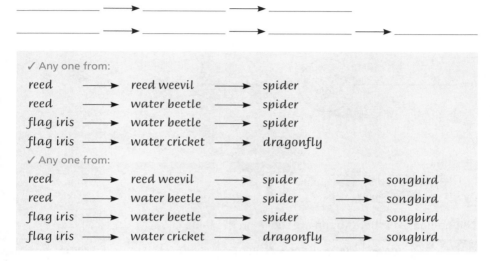

✓ Any one from:

reed ⟶	reed weevil ⟶	spider
reed ⟶	water beetle ⟶	spider
flag iris ⟶	water beetle ⟶	spider
flag iris ⟶	water cricket ⟶	dragonfly

✓ Any one from:

reed ⟶	reed weevil ⟶	spider ⟶	songbird
reed ⟶	water beetle ⟶	spider ⟶	songbird
flag iris ⟶	water beetle ⟶	spider ⟶	songbird
flag iris ⟶	water cricket ⟶	dragonfly ⟶	songbird

b) What would happen to numbers of primary consumers in the food web if numbers of water beetles declined? *(1 mark)*

They would increase ✓.

Surveying Populations

You will have carried out a required practical to measure the population size of a common species in a habitat. In all likelihood you used quadrats to measure populations of plant life.

As well as accomplishing this, you should be aware that ecologists use a variety of methods to assess distribution of species – including transects. These next questions explore these techniques.

Ecology

Example

Lichens are organisms that are sensitive to sulfur dioxide pollution.

Scientists wanted to investigate levels of pollution around an industrial area, so they carried out two line transects, as shown. At 200-metre intervals along each transect they counted the number of lichens growing on the nearest tree, as shown in the diagram.

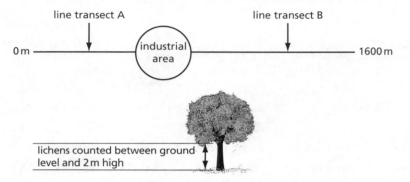

The number of lichens found are shown in the table.

Transect	200 m	400 m	600 m	800 m	1000 m	1200 m	1400 m	1600 m
A	0	0	3	4	5	8	9	8
B	2	5	8	9	9	7	8	9

a) Suggest why the results for the two transects are different. *(1 mark)*

> ✓ Any one from:
>
> the direction of the wind (affecting exposure to pollutants from industrial area)
>
> transect A may have passed over more roads, which could have caused increased pollution / sulfur dioxide levels.

b) What conclusion could you draw from the results of line transect A? *(1 mark)*

> The industrial area was producing sulfur dioxide pollution ✓.

c) Suggest why scientists did not count the number of lichens on the whole tree. *(2 marks)*

> The trees would have been different heights ✓.
> This would affect the reliability of the data ✓.

d) Suggest **one** factor scientists were unable to control that could affect the reliability of the results. *(1 mark)*

> ✓ Any one from:
>
> width of tree trunks
>
> distance of nearest tree to transect
>
> type of tree.

> This might be an example of a question where you are asked to apply knowledge you have acquired while carrying out your required ecology practical. (The specification states that you use sampling techniques to investigate the effect of a factor on the distribution of a species.)

Example

A class of students was asked to estimate the number of daisies on the school field. The field is 60 m by 90 m and has an area of 5400 m². They decided to use quadrats that were 1 m².

a) Which is the best way of using quadrats in this investigation?
Tick **one** box. *(1 mark)*

Place all the quadrats where there are lots of plants. ☐

Place all the quadrats randomly in the field. ☐

Place all the quadrats where daisies do not grow. ☐

> Place all the quadrats randomly in the field. ✓

Each student collected data by using ten quadrats. The results of one student, Shaun, are shown in the table below.

Quadrat	1	2	3	4	5	6	7	8	9	10
Number of daisies	5	2	1	0	4	5	2	0	6	3

b) i) Calculate the mean number of daisies per quadrat counted by Shaun. Show clearly how you worked out your answer. *(2 marks)*

> $(5 + 2 + 1 + 4 + 5 + 2 + 6 + 3) \div 10$ ✓ $= 28 \div 10 = 2.8$ ✓

ii) Calculate the median and mode of daisies per quadrat. *(4 marks)*

> Median – 3.5 ✓
>
> *Working – numbers arranged in order of magnitude: 1, 2, 2, 3, 4, 5, 5, 6.*
> *The middle two numbers are 3 and 4, therefore the median is 3.5 ✓.*
>
> 1 mark awarded if answer is incorrect but working is shown using the correct method.
>
> Mode – 3.5 ✓
>
> *Working – the values 2 and 5 appear most often, the mean of these two numbers is 3.5 ✓.*
>
> 1 mark awarded if answer is incorrect but working is shown using the correct method.

c) Another student, Bethany, calculated a mean of 2.3 daisies per quadrat from her results. Using Bethany's results, estimate the total number of daisies in the whole field by using the equation below. Show clearly how you work out your answer.

(2 marks)

estimated number of daisies in the field =
mean number of daisies per quadrat × number of quadrats that would fit into a field

> 2.3 × 5400 ✓ = 12 420 ✓
>
> 2 marks for a correct answer but, if incorrect, 1 mark will be given for correct working.

Predator-Prey Cycles

We have considered competition as a major factor influencing the distribution of organisms. Predation is the other main focus. The interaction between predators and prey is a popular area to be tested by examiners. Learn the basic principles, then apply these to the specific situation presented.

Example

The following information on the population of stoats and rabbits in a particular area was obtained over a period of ten years.

Year	1989	1990	1991	1992	1993	1994	1995	1996	1997	1998
No. of stoats	14	8	8	10	12	16	14	6	8	12
No. of rabbits	320	360	450	600	580	410	300	340	450	500

a) Plot these results onto the graph paper provided. *(3 marks)*

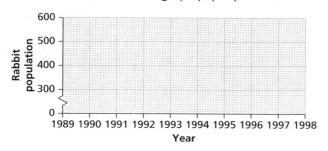

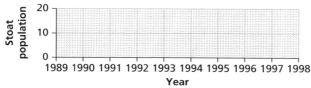

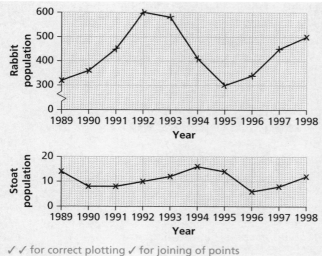

✓ ✓ for correct plotting ✓ for joining of points

b) Explain the reason for the variation in sizes of the stoat and rabbit populations.

(3 marks)

As the rabbit population decreases, there is less food (the rabbits) available for the stoats ✓.

So the stoat population decreases ✓.

Fewer rabbits are eaten, so the rabbit population increases ✓.

Ecology

You will notice in this graph that the points are connected by straight lines. This may be at odds with what you are taught in physics where a smooth curve is preferred. This method is used here because we cannot be certain of the values that lie between the points (it is not a direct relationship between numbers and time). In many physics experiments (and indeed chemistry) this is not the case and the values between the points are predictable. You could learn the answer to part **b)** by rote as it can be applied to any example of predator–prey cycles.

The Carbon Cycle

As well as the water cycle, you will need to understand how the element carbon moves through the living and non-living parts of the environment. You may wonder why carbon is special in this respect. As you will learn, the element forms the basis of every organic compound and therefore every biochemical process in living organisms. The main processes to look out for are respiration, photosynthesis and combustion.

Example

Carbon is recycled in the environment in a process called the carbon cycle. The main processes of the carbon cycle are shown here.

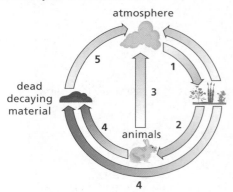

a) Name the process that occurs at stage 3 in the diagram. *(1 mark)*

> *Respiration ✓*

b) The UK government is planning to use fewer fossil-fuel-burning power stations in the future. How might this affect the carbon cycle? Use ideas about **combustion** and **fossil fuel** formation in your answer. *(2 marks)*

✓ ✓ Any two from:

fossil fuels represent a carbon 'sink' / they absorbed great quantities of carbon many millions of years ago from the atmosphere

combustion in power stations returns this carbon dioxide

less burning of fossil fuels cuts down on carbon emissions

alternative sources of energy may not add as much carbon dioxide to the atmosphere.

Biodiversity and Conservation

Most questions link the ideas of biodiversity and conservation. The main ideas are:

- biodiversity is a measure of the number of different species found in a given area or region
- high biodiversity is desirable from an ecological point of view
- adverse human impact reduces biodiversity
- carefully applied conservation measures can maintain biodiversity.

Example

Organic foods have become popular in recent years. They are grown without the use of pesticides and fertilisers.

A government report in 2007 showed that the production of some organic foods is more damaging to the environment than their non-organic equivalents.

However, supporters of organic farming claim that it is better than non-organic farming in conserving biodiversity and is better for the soil.

a) What is meant by the term biodiversity? *(1 mark)*

The range of species in a habitat ✓.

b) Why is it important to conserve biodiversity? *(1 mark)*

✓ Any one from:

they may have future uses

moral duty to maintain biodiversity

to maintain food webs and interactions between organisms in an environment.

c) State **two** features of non-organic farming that are thought to be damaging to the environment. *(2 marks)*

✓ ✓ Any two from:

application of pesticides

application of inorganic fertilisers

destruction of hedgerows

use of machinery

less humane enclosure of animals.

Example

In Ireland, four species of bumble bee are now endangered. Scientists are worried that numbers may become so low that they are inadequate to provide pollination to certain plants.

a) Explain how the disappearance of these four species of bumble bee might affect biodiversity as a whole. *(3 marks)*

Certain plant species / crops become scarce ✓.

Due to lack of pollinators ✓.

Other species that depend on these plants will be endangered / reduced in number; biodiversity falls ✓.

b) Threats to bumble bees are many: the possible effect of pesticides such as neonicotinoids, loss of suitable habitat, disappearance of wildflowers, new diseases.

Suggest **two** conservation measures that might help increase numbers of bumble bees in the wild. *(2 marks)*

✓ ✓ Any two from:

reduce usage of pesticides such as neonicotinoids

plant species that attract bumble bees (flowers, etc.)

farm bees in hives and release into the wild

disease prevention measures to act against foul-brood disease / isolate colonies / remove diseased colonies.

Waste Management

The AQA specification requires that you are able to describe the effects of specific pollution problems: sewage, fertilisers, toxic chemicals, smoke and acidic gases in the air, landfill. The next question is in context of acid rain and its effects on plants.

Example

Dylan and Molly investigated the effect of sulfur dioxide on the germination of cress seeds. The diagram shows their apparatus.

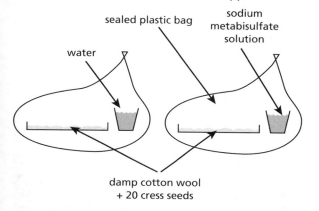

- Sodium metabisulfate solution gives off sulfur dioxide.

- Both bags were left in a warm laboratory for five days.

a) What was the independent variable in the investigation? *(1 mark)*

The presence of water or sodium metabisulfate ✓.

b) Suggest the main reason for using sealed plastic bags. *(1 mark)*

To keep the gases in ✓.

Dylan and Molly counted the number of seeds that had germinated after five days. Their results are shown in the table below.

	Number of germinated seeds
Water	18
Sodium metabisulfate	12

c) i) What conclusion can Dylan and Molly draw from their results? *(1 mark)*

That sodium metabisulfate affects (decreases) the germination of cress seeds ✓.

ii) Are their results reliable? Explain your answer. *(2 marks)*

No ✓. *They need to repeat the experiment to increase the reliability of their results* ✓.

d) Suggest how acid rain might affect the biodiversity of plants. *(2 marks)*

> Plant biodiversity lowered (no marks) due to plants being killed through
> ✓ ✓ Any two from:
> effects of acid on bark
> leaves being bleached / damaged
> soil nutrients being affected
> pH of soil being lowered.

Land Use

Example

The picture shows a peat bog.

Peat is formed by the compression of plant remains over millions of years.

Peat bogs are ancient habitats that can act as carbon sinks, which means they can store and 'lock up' carbon.

a) Explain how a peat bog acts in this way. In your answer use ideas about the carbon cycle and photosynthesis. *(3 marks)*

> ✓ ✓ ✓ Any three from:
> photosynthesis absorbs carbon dioxide from the atmosphere
> plants die but do not decay
> fossilisation of plant remains
> peat containing carbon compounds remains in the ground.

b) Peat bogs are declining in area across the world due to extraction for gardening purposes. Explain the effects of this habitat destruction on:

- biodiversity
- climate change. *(3 marks)*

> Biodiversity lowered because habitats are destroyed for a wide range of organisms that rely on them ✓. Extraction of peat releases carbon dioxide ✓, which increases global temperatures / hastens climate change ✓.

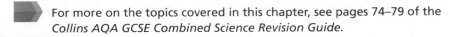

For more on the topics covered in this chapter, see pages 74–79 of the *Collins AQA GCSE Combined Science Revision Guide*.

8

Atomic Structure and the Periodic Table

Atoms, Elements and Compounds

The terms **atoms**, **elements** and **compounds** are commonly used in chemistry exams and so it is important that you know the meaning of each of them, as this will help you focus on what the question is asking you to do.

Example

State the name of the element and the number of each atom of that element present in the compound with the formula $Ca(NO_3)_2$. *(2 marks)*

Element	No. of atoms present
Calcium	1
Nitrogen	2
Oxygen	6
✓	✓

Use the Periodic Table (see page 304) to find the names of the elements present. The Periodic Table is provided in both papers.

A common error is not realising that the number outside of the brackets multiplies the number of atoms inside the brackets. There are three oxygen atoms inside the brackets and this is multiplied by two because there is a '2' outside of the brackets. There is only one atom of calcium present because there are no numbers attached to it.

Example

Define the word 'compound'. *(2 marks)*

A compound contains two or more different elements ✓ chemically bonded together ✓ in fixed proportions.

If there is only one element bonded to itself, e.g. O_2, then it is an element and not a compound.

If the atoms are not chemically bonded together, then they are a **mixture** and not a compound.

All compounds contain atoms combined in a fixed proportion, e.g. H_2O consists of two atoms of hydrogen chemically bonded to one atom of oxygen.

Example

Which statement about atoms, elements and compounds is **false**? Tick **one** box. *(1 mark)*

There are about 50 different elements. ☐

Compounds are formed from elements by chemical reactions. ☐

Compounds can only be separated into elements by chemical reactions. ☐

All substances are made from atoms. ☐

Multiple choice questions are used to test knowledge as well as understanding. This question is testing knowledge. At first glance, all options given might seem reasonable but this particular question relies on you knowing that there are approximately 100 different elements. This fact is stated in the specification and therefore you are expected to know it!

There are about 50 different elements. ✓

Mixtures

Mixtures differ from compounds in that they consist of two or more elements (or compounds) not chemically combined together. Salt dissolved in water is a mixture. As you can dissolve different amounts of salt in the same volume of water, they are not together in a fixed quantity. Mixtures can be easily separated by physical processes such as **filtration, crystallisation, simple distillation, fractional distillation** and **chromatography**.

You should know these different methods of separation and be able to decide which method is appropriate for separating a given mixture. You could also be asked to draw and / or label appropriate apparatus to carry out these methods of separating mixtures.

Example

Draw and label the apparatus that could be used in a laboratory to obtain **pure** water from a mixture of salty water.   *(4 marks)*

You should first decide which is the appropriate method of separation to use. You should then draw and label the apparatus you choose to use. Ensure that your diagram answers the question, i.e. makes it clear how pure water is to be obtained.

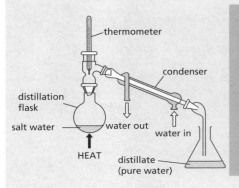

You do not need to include the bench and clamps, etc. The examiner will appreciate that your apparatus isn't just floating in mid-air!

Alternative names for apparatus are acceptable, for example the distillation flask can be called a 'round-bottomed flask'. The condenser may be referred to as a water-cooled condenser or Liebig condenser.

The thermometer is optional but it is helpful to know the temperature of the substance entering the condenser.

Salt water in flask connected to a condenser ✓   Collection of pure water ✓

Correct water flow through the condenser ✓   Correct labels ✓

The Development of the Model of the Atom

There will be questions assessing 'working scientifically' and this topic is one that considers how scientific methods and theories develop as well as the use of models to represent theories and ideas within science. Over time, scientists have considered atoms to be tiny indivisible spheres, having a structure modelled on a 'plum pudding', through to the nuclear model we use today.

Example

In 1864, atoms were thought to be **particles** that could not be divided up into smaller particles. By 1898, the **electron** had been discovered and the plum-pudding model of an atom was proposed.

The diagram below shows the plum-pudding model of an atom of carbon and the nuclear model of an atom of carbon.

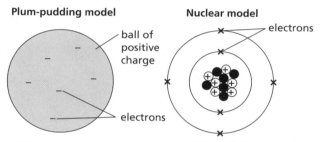

a)   Describe the differences between the plum-pudding and nuclear models.   *(3 marks)*

Student A

Both models of the atom contain positive charges and electrons. In the plum-pudding model, the electrons are randomly placed in the positive charge and in the nuclear model the electrons surround the positive charge ✓ (M2).

Student B

The plum-pudding model considers the atom to be a ball of positive charge with negatively charged electrons embedded within it. The nuclear model considers the positive charge within an atom to be in particles called **protons** located in the centre of an atom in a region known as the nucleus ✓ (M1). In addition, the nucleus also contains neutral particles called **neutrons** (shown as black dots) in the diagram ✓ (M3). In the nuclear model the electrons are not embedded within the positive charge but are arranged in shells that surround the central nucleus ✓ (M2).

Mark scheme

M1 Ball of positive charge in plum-pudding model and positive charge in centre of atom / nucleus in the nuclear model

M2 Electrons embedded in positive charge in plum-pudding model and around the nucleus / in shells in the nuclear model

M3 No neutrons in plum-pudding model and presence of neutrons in the nuclear model

The two command words in this question are 'describe' and 'explain'. Describe means to recall some facts, events or processes in an accurate way. When a question asks you to 'explain', it is asking you to make something clear or state the reasons for something happening.

b) Explain how the work of different scientists contributed to the development of the plum-pudding model. *(4 marks)*

Student A

Niels Bohr stated that electrons orbit the nucleus of an atom at specific distances and that is why the electrons in the nuclear model surround the nucleus in shells ✓ (M5).

Student B

The results from the alpha particle scattering experiment led to the idea that the mass of an atom was concentrated at the centre of an atom in the region we now call the nucleus and that the nucleus was charged ✓ (M4). Niels Bohr adapted this nuclear model by suggesting that electrons orbit the nucleus at specific distances ✓ (M5). His theoretical calculations agreed with experimental observations. Later experiments led to the idea that the positive charge of any nucleus could be subdivided into a whole number of small particles, each of which has the same amount of charge ✓ (M6). These particles are now called protons. Subsequent experimental work carried out by James Chadwick provided the evidence to show the existence of neutrons within the nucleus of an atom ✓ (M7).

Mark scheme

M4 Reference to the alpha particle scattering experiment and the conclusion from this that the mass of an atom was concentrated at the centre of the atom

M5 Statement that the work of (Niels) Bohr suggested that electrons obit the nucleus at specific distances / are found in shells (or energy levels) around the nucleus

M6 Reference to experimental work showing positive charge in individual particles (protons)

M7 Mention of the work of James Chadwick leading to the idea of neutrons.

Student A would be awarded two marks across parts **a)** and **b)**. They haven't quite given enough detail to be awarded M1 as they haven't indicated that the positive charge is contained within the centre of the atom / within the nucleus. They have just referred to the diagram rather than accurately describing the differences.

Student A makes no reference to neutrons or the alpha particle scattering experiment and therefore does not score M3 or M4. They also do not reference the experimental work that led to the idea of the positive charge being contained in individual particles and so are not awarded M6. To be awarded M7, the contribution of James Chadwick must be mentioned.

Student B would be awarded all seven marks.

Atoms

Knowing the numbers of protons, neutrons and electrons is key to understanding atomic structure. This can then help with other topics such as chemical reactions and bonding.

The **atomic number** and **mass number** allow us to quickly determine the number of each sub-atomic particle in an atom.

You also need to know the mass and charge of the sub-atomic particles, as well as having an appreciation of the size of atoms.

Example

An atom of an unknown element X can be represented as shown below.

$$^{40}_{18}X$$

Atomic Structure and the Periodic Table

a) Calculate the number of protons, neutrons and electrons in an atom of X. *(3 marks)*

Protons = 18 ✓	The lower number in the representation is the atomic number. The atomic number tells you the number of protons in an atom.
Electrons = 18 ✓	The number of protons in an atom is equal to the number of electrons.
Neutrons = 22 ✓	The upper number in the representation of element X is the mass number. The number of neutrons in an atom is calculated by subtracting the atomic number from the mass number, i.e. 40 − 18 = 22.

b) State the name of element X. *(1 mark)*

The atomic number of an element tells you the identity of that element. Look at the Periodic Table to find the element with atomic number 18.

Element X = Argon ✓

Example

Complete the following sentences. Tick **one** box for each sentence. *(2 marks)*

Atoms are very small, having a radius of about _____.

0.1 nm	☐	10 nm	☐
1 nm	☐	100 nm	☐

The radius of a nucleus is approximately _____ of that of the atom.

$\frac{1}{100}$	☐	$\frac{1}{10000}$	☐
$\frac{1}{1000}$	☐	$\frac{1}{1000000}$	☐

$0.1\,nm$ ✓ and $\frac{1}{10000}\,nm$ ✓.

This question relies on you knowing these facts.

Relative Atomic Mass

Many elements exist as **isotopes** (atoms of the same element that have the same number of protons but different numbers of neutrons). The **relative atomic mass** of an element takes into account the different abundances (amounts) and masses of each isotope.

Example

In a sample of copper, 69.1% of the atoms have a mass number of 63 and the remainder have a mass number of 65.

Use the above information to calculate the relative atomic mass (RAM) of copper. Give your answer to three significant figures. *(3 marks)*

The amount of copper-65 is

$100 - 69.1 = 30.9\%$ ✓

$RAM = \frac{(69.1 \times 63) + (30.9 \times 65)}{100}$ ✓

$RAM = 63.6$ (3 s.f.) ✓

As there are only two isotopes of copper (copper-63 and copper-65), the total amount of them must add up to 100%. As you are calculating the average mass of isotopes that have a mass of either 63 or 65, your answer will be somewhere between these two values.

It is a good idea to write 3 s.f. on the answer line before you start your calculation. Then you are not likely to forget to give your answer to this number of significant figures.

Electronic Structure

Knowing the electronic structure of atoms is fundamental in understanding how atoms react and bond together.

Example

An atom of element Z has 16 protons. Complete the diagram below to show the **electronic configuration** of element Z. *(2 marks)*

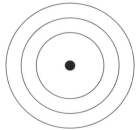

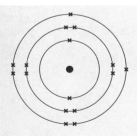

There are 16 electrons (shown as crosses in this diagram) because the atomic number of element Z is 16. The correct electronic structure is therefore to have two electrons in the first energy level / shell, eight in the second energy level and six in the outer energy level.

Total of 16 electrons ✓
Correct number of electrons in each energy level ✓

The Periodic Table and its Development

The Periodic Table is arranged in a way that provides useful information. The elements are arranged in order of increasing atomic number, and elements with similar chemical properties are in the same group. The group number tells you how many electrons are in the outer shell of the atom, e.g. all elements in Group 7 (the halogens) have seven electrons in the outer shell of their atoms.

Many scientists have worked on producing periodic tables but **Mendeleev** is considered to be largely responsible for the structure of the Periodic Table we use today.

Example

In 1869 Dmitri Mendeleev produced an early version of the Periodic Table.

a) Complete each sentence. Tick **one** box for each sentence.   *(3 marks)*

i) Mendeleev first arranged the elements in order of their

date of discovery. ☐

atomic weight. ☐

electronic configuration. ☐

reactivity. ☐

atomic weight. ✓

ii) Mendeleev also placed the elements with similar properties in columns called

periods. ☐

groups. ☐

shells. ☐

reactivity series. ☐

groups. ✓

iii) When elements did not fit Mendeleev's pattern he

ignored the element. ☐

left a gap. ☐

put the element in a group of 'odd' elements. ☐

changed the order. ☐

left a gap. ✓

This question requires you to understand the structure of the Periodic Table as well as to have an appreciation of the contribution of Mendeleev to its development.

b) Mendeleev's 1869 Periodic Table did not include the **noble gases**. Suggest why this group was not included. *(1 mark)*

The command word 'suggest' is used when you are required to apply your knowledge and understanding to a new situation.

The noble gases (due to their lack of reactivity) had not been discovered ✓.

Example

a) Use the correct word from the box to complete each sentence. Each word can be used once, more than once, or not at all. *(2 marks)*

| electrons | molecules | neutrons | protons |

Questions like this are variations of the standard multiple-choice type.

i) Elements in the modern Periodic Table are arranged in order of the number of _____ in their nucleus.

protons ✓

ii) Elements in the same group have the same number of _____ in their outer energy level.

electrons ✓

Atomic Structure and the Periodic Table

Groups 0, 1 and 7

Elements in Groups 0, 1 and 7 are characterised by having similar chemical properties and by displaying a trend in physical properties such as boiling point. The **transition metals** are located between Groups 2 and 3 and their properties are frequently compared with the Group 1 metals.

Example

Sodium and lithium are both Group 1 elements. They are stored in oil to prevent them reacting with oxygen in the air.

a) Write a balanced chemical **equation** for the reaction of sodium with oxygen to form sodium oxide. *(2 marks)*

$$4Na_{(s)} + O_{2(g)} \rightarrow 2Na_2O_{(s)}$$

Correct formulae ✓

Correctly balanced equation ✓

A chemical equation means using formulae. State symbols will be asked for if required and, whilst they aren't required in this question, they are shown anyway. It is a good idea to write state symbols where possible so that you develop this habit. You will not be penalised for getting them wrong unless the question asks you to include them.

b) Explain why sodium is more reactive than lithium. *(3 marks)*

Sodium atoms are larger than lithium atoms ✓.

This means the outer electron in sodium is further away from the nucleus / less attraction between the nucleus and outer shell electron ✓.

Therefore the outer electron in sodium is more easily lost ✓, hence sodium is more reactive than lithium.

This is a good question for assessing your understanding of the reactivity of the Group 1 elements. You may find that a diagram (in this case showing the different sizes of atoms of sodium and lithium) helps to explain and focus your answer. Many student responses to questions such as this one lack a logical approach, so try to ensure that each point links to the next one.

Many students often use the word 'it' when answering questions. Either try to avoid the use of this word or make it very clear what 'it' is referring to!

Example

Chlorine and bromine are both in Group 7. Elements in Group 7 are known as the **halogens**.

a) Which one of the following statements about the halogens is **false**?

Tick **one** box.   *(1 mark)*

The halogens consist of molecules made of pairs of atoms.   ☐

The reactivity of the halogens increases further down the group.   ☐

Relative molecular mass increases further down the group.   ☐

Bromine has a higher boiling point than chlorine.   ☐

The reactivity of the halogens increases further down the group. ✓	You should know the physical characteristics and properties of the elements in Groups 0, 1 and 7 and the transition metals.

b) Complete the word equation for the reaction between chlorine and sodium bromide.   *(1 mark)*

chlorine + sodium bromide → _____ + _____

When answering questions involving equations, try to work out what type of reaction is taking place. The type of reaction occurring in this question is a **displacement reaction**. In displacement reactions of halogens, the more reactive halogen replaces the less reactive halogen in a compound.

*The correct **products** are bromine and sodium chloride* ✓.	In this question both products are required to be awarded the mark.

c) **HT** The reaction between bromine and iodide **ions** can be represented by an **ionic equation**. Complete the following ionic equation.   *(2 marks)*

$Br_2 + 2I^- \rightarrow$ _____ + _____

$2Br^-$ ✓ *and I_2* ✓	Bromine is more reactive than iodine and so bromine atoms are able to gain electrons from the iodide ions. This results in iodine molecules and bromide ions being formed.

Example

Sodium reacts with fluorine to form the compound sodium fluoride (NaF). Sodium fluoride is a solid at room temperature.

a) Write a chemical equation for the reaction between sodium and fluorine. Include state symbols. *(2 marks)*

Remember that some elements are diatomic, i.e. molecules consisting of two atoms. The diatomic elements are hydrogen, oxygen, nitrogen and the Group 7 elements. Being able to work out chemical formulae is a very important skill in chemistry.

$$2Na_{(s)} + F_{2(g)} \rightarrow 2NaF_{(s)}$$ Correct equation ✓   Correct state symbols ✓

b) Draw a diagram to show the electronic structure of an atom of sodium before it has reacted with fluorine. *(1 mark)*

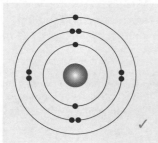

This question is asking you to draw the electronic structure of an atom of sodium. Remember that the number of electrons in an atom is the same as the atomic number of that element. Two electrons occupy the first energy level and up to eight in the others.

You can use crosses or dots to represent the electrons. However, be consistent and ensure all of an atom's electrons are represented in the same way.

c) Explain how the electronic structure of an atom of fluorine is different **after** it has reacted with sodium. *(1 mark)*

The fluorine atom will now have a full outer energy level of electrons ✓ (having gained an extra electron from the sodium atom).

Atoms react in order to gain a full outer energy level of electrons. Fluorine atoms have seven electrons in their outer energy level and so they gain one more electron (from the atoms they are reacting with) to achieve a full outer energy level.

The Group 7 element astatine also reacts with sodium. The compound formed is called sodium astatide.

d) Predict the formula of sodium astatide. *(1 mark)*

Elements in the same group have similar properties. The question tells you that sodium fluoride has the formula NaF, i.e. one atom of sodium combines with one atom of fluorine. As astatine is in the same group as fluorine, then it would be expected that astatine would react with sodium in the same ratio.

NaAt ✓

e) Will fluorine or astatine react more vigorously with sodium? Explain your answer. *(4 marks)*

The first part of the question is asking you to recall the trend in reactivity of the Group 7 elements. The explanation for this reactivity is very similar to the explanation for the reactivity of the Group 1 elements: the difference being that atoms of Group 1 elements 'want' to lose an electron, whereas atoms of Group 7 elements 'want' to gain an electron.

Fluorine ✓

Fluorine atoms are smaller than astatine atoms ✓.

This means the outer energy level is closer to the nucleus in an atom of fluorine ✓.

Therefore atoms of fluorine have a greater attraction for incoming electrons, meaning that they are more reactive ✓.

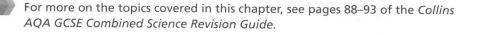

For more on the topics covered in this chapter, see pages 88–93 of the *Collins AQA GCSE Combined Science Revision Guide*.

9

Bonding, Structure and the Properties of Matter

The way that atoms bond together and the resulting structures help us to explain many of the chemical and physical properties of different substances. The three strongest chemical bonds are ionic, covalent and metallic. Scientists use their knowledge of structure and bonding to design new materials to solve problems and improve our lives.

Ionic Bonding

Ionic bonding occurs between metals and non-metals. Electrons from the outer shell (energy level) of metal atoms are transferred to the outer shell (energy level) of non-metal atoms. The resulting charged atoms are known as ions. The actual bond is formed by the attraction of the oppositely charged ions. Dot and cross diagrams are used to show the transfer of electrons during the formation of an **ionic bond**.

Example

Draw a dot and cross diagram to show the formation of the ionic bond in sodium chloride. Include the charges on the ions formed. *(3 marks)*

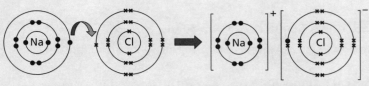

Correct electronic configuration of the starting atoms ✓

Correct electronic configuration of the resulting ions ✓

Correct charges on the resulting ions ✓

The arrow showing the transfer of the electron from the sodium atom to the chlorine atom is not necessary but is shown here to help explain what is happening when sodium and chlorine react.

It does not matter which atom's electrons are shown as dots or crosses.

You can show the sodium ion having either eight electrons or without any electrons in the outer shell.

Example

The electronic configurations of atoms of calcium and fluorine are:

Ca   2,8,8,2

F   2,7

Describe the changes in the electronic configurations of calcium and fluorine when these atoms react to form calcium fluoride. *(3 marks)*

Ca atoms lose two electrons ✓.

(Two) F atoms gain an electron ✓.

Ca becomes 2,8,8

and F becomes 2,8 ✓.

This question is very similar in content to the previous one. However, the emphasis in this question is on electronic configurations. Again, a diagram might help to focus and explain your answer but a diagram should not be given instead of a written description.

Example

Metal X forms a chloride with the formula XCl_2.

a) Give the formula of the ion of X. *(1 mark)*

X^{2+} ✓

Think about the ion that chlorine atoms will form (chlorine is in Group 7). There are two chloride ions present and the charge on X must cancel out the charge on these two ions.

b) State and explain which group of the Periodic Table element X is in. *(2 marks)*

Group 2 ✓ as an atom of X contains two electrons in its outer energy level (shell) ✓.

If X forms a 2+ ion (answer from part **a)**) then an atom of X must have two electrons in its outer energy level / shell. This means it is in Group 2.

c) Suggest the type of bonding present in XCl_2. *(1 mark)*

Ionic ✓

X is a metal and chlorine is a non-metal. Ionic bonding occurs when a metal reacts with a non-metal.

Ionic Compounds

Compounds containing ionic bonding form giant ionic structures. The ions are held together (in a structure known as a lattice) by strong **electrostatic** forces of attraction that act in all directions between oppositely charged ions.

Example

The two diagrams below represent two ways that the 3-D structure of sodium chloride can be shown.

Ball and stick model

Giant lattice

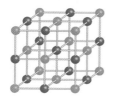

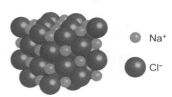

Na⁺

Cl⁻

Bonding, Structure and the Properties of Matter

a) For one of the diagrams, describe **one** limitation of representing giant ionic compounds in this way. *(1 mark)*

> There are various models used to represent bonding and structures. As these are models, they have limitations. Ensure that you understand these limitations.

> ✓ One from:
>
> *the ball and stick model implies that the ions are physically held apart /*
>
> *both models show the ions as perfect spheres /*
>
> *neither model shows the exact numbers of ions present in an ionic lattice.*

b) Use the 2-D diagram below to state the **empirical formula** of sodium chloride. Explain your answer. *(2 marks)*

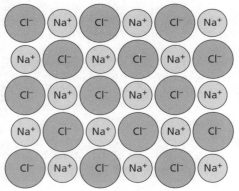

> *The empirical formula is NaCl ✓ because there are an equal number of sodium ions and chloride ions ✓.*

> The empirical formula is 'the simplest whole number ratio of each atom present in a compound'. Whilst you may know the formula of sodium chloride to be NaCl, it is important to be able to recognise why this is also the empirical formula in case you are given an example that you are not familiar with.

> By counting up the number of each sodium ion and chloride ion in the diagram, you can see that there is an equal number of each (15). As they are present in a 1:1 ratio then the empirical formula is NaCl.

Covalent Bonding

A **covalent bond** is a shared pair of electrons between two non-metal atoms. Substances with covalent bonds can exist as either small molecules, very large molecules such as **polymers** or giant covalent (macromoleular) substances.

Example

Draw a dot and cross diagram to show the bonding in ammonia (NH_3). *(2 marks)*

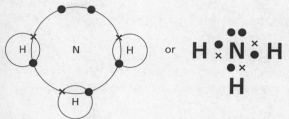

Three atoms of hydrogen attached to nitrogen with electrons represented as a dot and a cross ✓

Two non-bonded electrons in the outer shell of nitrogen ✓

Questions asking you to draw dot and cross diagrams are very common. It can be helpful to draw the outer shell of each atom before it has bonded to see what is going to happen. Remember that a covalent bond involves two electrons – one from each atom.

This diagram (right) is also perfectly acceptable as an alternative to the first diagram.

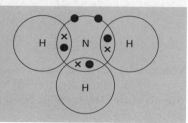

Example

Hydrazine (N_2H_4) has this structure:

$$\begin{array}{c} H \quad\quad H \\ \backslash \quad\quad / \\ N - N \\ / \quad\quad \backslash \\ H \quad\quad H \end{array}$$

Complete the diagram below to show the bonding in hydrazine. You only need to show outer shell electrons. *(3 marks)*

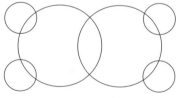

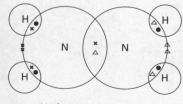

Two atoms of hydrogen attached to each nitrogen with a pair of electrons between them ✓

A pair of electrons between the two nitrogen atoms ✓

A pair / two electrons in the outer shell of each nitrogen atom ✓

● = H electron
✗ = electron from 1st N atom
△ = electron from 2nd N atom

This question is also asking you to draw a dot and cross diagram but for an unfamiliar compound. The structure shown at the beginning of the question tells you that there is one covalent bond (i.e. two electrons) between each atom. Nitrogen has five electrons in its outer shell meaning that after forming three bonds there will be two electrons left.

As long as there are the correct number of electrons in the right places the examiner will not be worried about whether they are shown as dots or crosses (triangles have also been used in the answer to help you see which atom the electrons belong to).

Questions like this will usually tell you to show only the outer shell electrons.

Metallic Bonding

In a metal, each atom loses its outer shell electrons leaving a lattice (regular repeating pattern) of cations surrounded by **delocalised** electrons that are free to move through the structure. The **metallic bond** is the attraction between the cations and these delocalised electrons.

Example

Thallium is a metal in Group 3 of the Periodic Table.

Describe the structure and bonding in thallium. *(3 marks)*

Lattice / regular arrangement / giant structure ✓.

Positive ions / cations ✓.

Surrounded by delocalised electrons ✓.

Thallium may not be well known to you but you are told (and should be able to confirm from its position in the Periodic Table) that it is a metal.

A labelled diagram such as the one below may help to structure and focus your answer but it should not be used instead of a description.

delocalised electrons from outer shell of metal atoms

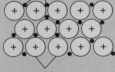

metal ions

States of Matter

The particle model can be used to represent the three states of **matter**. Melting and freezing take place at the **melting point** whereas boiling and condensing take place at the boiling point.

Example

A sample of water is heated from a liquid at 40°C to a gas at 110°C.

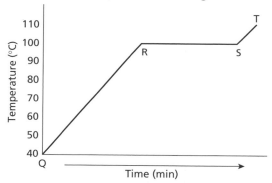

a) On the heating curve diagram, label each of the following regions:

Liquid, only

Gas, only

Change of state *(1 mark)*

Liquid only – the line between Q and R labelled Gas only – the line between S and T labelled Change of state – the line between R and S labelled ✓.	When attempting data analysis questions, it is worth spending time thinking about how the data was generated and where possible picturing the experiment that was carried out to generate the data.

b) For section QR of the graph, state what is happening, in terms of energy and movement, to the water molecules as the water is heated. *(2 marks)*

The particles gain kinetic energy ✓ and move more quickly ✓.	The water is being heated and the temperature is increasing. So what is happening to the water molecules?

c) For section RS of the graph, state what is happening to the water molecules and explain why there is no change in temperature. *(2 marks)*

> The bonds between the water molecules are breaking ✓ and this process requires energy, which is why there is no change in temperature ✓.

> The water is being heated but the temperature is not increasing. So what is happening to the energy that is being transferred to the water?

Example

The particle model can be used to represent the arrangement of particles in solids, liquids and gases. The particle model representation of a liquid is shown below.

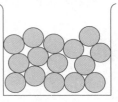

a) In the space below, show the arrangement of particles in a solid. *(1 mark)*

solid

Particles closer together than in a liquid and in a regular arrangement ✓

> You should know the names of all the changes of state and the arrangement of particles in the three states of matter.

b) What name is given to the process of changing from a liquid to a solid? *(1 mark)*

> Freezing ✓

c) **HT** State **one** limitation of the particle model. *(1 mark)*

✓ One from:

*in the particle model no forces
between particles are shown /*

*all of the particles are represented
as spheres /*

all the spheres are shown as being solid.

Remember that where models are
used in chemistry to show / visualise a
point there will always be limitations
to that model.

Properties of Materials

Different structures have different properties that can be explained by considering
the bonding present in the substance.

Example

Which one of the following substances **does not** conduct electricity?

Tick **one** box. *(1 mark)*

molten sodium ☐

molten lithium fluoride ☐

molten sulfur ☐

graphene ☐

molten sulfur ✓

Think about the structure that each of these substances has:
sodium – metallic
lithium fluoride – ionic
sulfur – simple molecular
graphene – giant covalent.
Simple molecular structures do not conduct electricity. Giant covalent structures usually
don't conduct electricity but graphene is one of the exceptions that you need to know!

Example

Which one of the following substances has the **lowest** boiling point?

Tick **one** box. *(1 mark)*

magnesium ☐   carbon dioxide ☐

lithium fluoride ☐   graphite ☐

Bonding, Structure and the Properties of Matter

carbon dioxide ✓

Simple molecular structures will (almost always) have lower boiling (or melting) points than the other types of structure.

Example

The table below shows some properties of four substances: A, B, C and D.

Substance	Melting point in °C	Boiling point in °C	Conducts electricity when solid?	Conducts electricity when molten?
A	801	1413	No	Yes
B	−101	−35	No	No
C	1063	2970	Yes	Yes
D	3549	4830	No	No

Use the information in the table to identify the substance that

a) has a simple molecular structure.

b) contains cations **and** anions.

c) is a transition metal.

d) is diamond. *(4 marks)*

a) B ✓

b) A ✓

c) C ✓

d) D ✓

Tables of data such as this one require you to apply your knowledge. In this case it involves relating the properties as given in the headings to the structure and bonding of different materials.

a) Substance B is a gas at room temperature (20°C). All gases at room temperature have a simple molecular structure.

b) Substance A does not conduct electricity when it is a solid but it does when **molten**. This means that it is an ionic substance and therefore contains cations **and** anions.

c) Transition metals have relatively high (when compared with Group 1 metals) melting points. Metals also conduct electricity when solid or molten.

d) **Diamond** (a substance whose structure you need to know) has a giant covalent / macromolecular structure and so it has a high melting and boiling point. Diamond does not conduct electricity (i.e. it is an electrical insulator).

Example

Carbon dioxide and **silicon dioxide** have different structures.

a) What type of bonding is present in both carbon dioxide and silicon dioxide? *(1 mark)*

> Covalent ✓
>
> Remember that covalent bonding occurs between non-metal atoms.

b) The diagram below shows a dot and cross diagram for carbon dioxide.

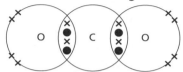

i) How many bonds are there between the carbon atom and each oxygen atom? *(1 mark)*

> Two ✓
>
> A covalent bond is a shared pair of electrons.

ii) Does carbon dioxide have a simple molecular or giant covalent (macromolecular) structure? *(1 mark)*

> Simple molecular ✓
>
> Carbon dioxide has three atoms per molecule. This number is fixed, i.e. you can't bond any more atoms to it and so it has a simple molecular structure.

c) The structure of silicon dioxide is shown below.

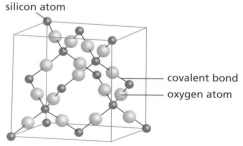

silicon atom

covalent bond

oxygen atom

i) Does silicon dioxide have a simple molecular or giant covalent (macromolecular) structure? *(1 mark)*

> Giant covalent / macromolecular ✓
>
> By looking at the diagram you can see that it could be continued. This diagram could theoretically go on for ever and so the structure is classified as giant.

ii) Structurally, what is the difference between a simple molecular structure and a giant covalent (macromolecular) structure? *(2 marks)*

> Simple molecular structures consist of small molecules containing a fixed number of atoms ✓ whereas giant structures have variable numbers of atoms in them / their structures have no defined boundaries ✓.
>
> This question requires you to communicate the explanations for the two previous answers.

d) Would you expect silicon dioxide to have a higher or lower melting point than carbon dioxide? Explain your answer. *(4 marks)*

> Silicon dioxide has a higher melting point than carbon dioxide ✓. The bonds broken when simple molecular structures are melted are **intermolecular** forces ✓. The bonds broken when giant covalent / macromolecular structures are melted are covalent bonds ✓. Covalent bonds are much stronger than intermolecular forces and so require much more energy to break them, hence a higher melting point ✓.
>
> Questions on melting / boiling point always require you to think of the bonds / forces that hold the structure together.

Example

Iron is a metal often used in engineering. The strength of iron can be increased by mixing it with other metals such as manganese or chromium.

a) State the bonding present in iron. *(1 mark)*

> Metallic ✓
>
> Iron is a metal so it has metallic bonding.

b) Explain why iron conducts electricity. *(2 marks)*

> Iron has delocalised electrons ✓ that are able to flow / move throughout the structure ✓.

This question asks you to explain a property of a material based on its structure.

c) Explain why metals such as iron are **malleable**, i.e. easily bent and shaped. *(2 marks)*

> The layers (of cations / atoms) ✓ are able to slide over each other ✓.

Again, the property is related to the structure.

d) What name is given to a mixture of metals such as iron and manganese? *(1 mark)*

> Alloys ✓

Alloys are formed when a metal is mixed with another substance (usually a metal).

e) Explain why a mixture of iron and manganese is harder than pure iron. *(2 marks)*

> The different atoms in an alloy are of different sizes ✓, meaning that the layers of atoms / cations are distorted and not able to slide over each other ✓.

This question highlights how changing the structure of a substance can affect its properties.

f) Will a mixture of iron and manganese conduct electricity? Explain your answer. *(2 marks)*

> Yes ✓. The mixture of metals will still have a metallic structure / there are still delocalised electrons ✓.

As this alloy is a mixture of two metals it will still conduct electricity as mobile delocalised electrons will still be present.

Example

a) What type of substance is represented by the structure below? *(1 mark)*

$$\left(\begin{array}{cc} H & H \\ | & | \\ -C - C - \\ | & | \\ H & H \end{array}\right)_n$$

Bonding, Structure and the Properties of Matter

> Polymer or poly(ethene) ✓
>
> The diagram shows the repeat unit of a polymer; in this case the polymer is poly(ethene).

b) What type of bonds hold the atoms together in this substance? *(1 mark)*

> Covalent bonds ✓
>
> Any diagram showing a 'stick' between atoms is showing a covalent bond.

c) Explain why you would expect this substance to be a solid at room temperature. *(2 marks)*

> The intermolecular forces between (polymer) molecules / chains ✓ are strong ✓.
>
> As polymer molecules are very large there will be lots of intermolecular forces between them. Even though intermolecular forces are relatively weak, the large number of them results in a lot of energy being required to overcome them. For this reason, polymers are solids at room temperature.

Example

The table below includes information about the electrical conductivity of sodium, chlorine and sodium chloride.

Substance	Electrical conductor?
Sodium	Yes
Chlorine	No
Sodium chloride	Only when molten or **aqueous**

With reference to the structure and bonding, explain the property of electrical conductivity of each substance. *(6 marks)*

> ✓ ✓ ✓ ✓ ✓ ✓ Any six points from a possible eight but no more than two from each substance.
>
> Sodium has a giant metallic structure ✓, meaning that there are delocalised electrons ✓ that are free to move throughout the structure and carry charge ✓.
>
> Chlorine contains covalent bonds and has a simple molecular structure ✓. There are no ions or delocalised electrons in chlorine and so it is unable to conduct electricity ✓.
>
> Sodium chloride has a giant ionic structure ✓. In the solid state, the ions are in fixed positions and not able to move ✓. However, in the liquid state (i.e. molten or aqueous) the ions are free to move and carry the charge ✓.

Two common misconceptions about electrical conductivity are:

1) 'Electrical conductivity only involves electrons.'

Charged particles (i.e. electrons and ions) can both carry charge and hence conduct electricity. In a metal it is the delocalised electrons that carry charge. In a molten or aqueous ionic compound it is the ions that carry the charge.

2) 'Ionic compounds conduct electricity because they contain a metal and as metals conduct electricity so must ionic compounds.'

Ionic compounds do contain metals but this is not why they conduct electricity. Ionic compounds contain ions and not delocalised electrons. When these ions are able to move electricity is conducted.

Structure and Bonding of Carbon

Carbon has many different structures including **graphite**, diamond, **fullerenes** and graphene. The diversity of structures and properties that carbon has provide us with many useful everyday materials.

Example

The diagrams show three different forms of carbon.

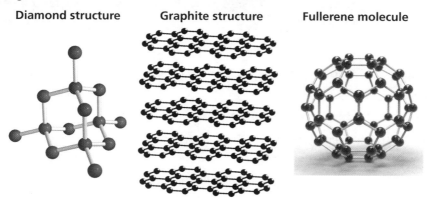

Diamond structure   **Graphite structure**   **Fullerene molecule**

a) Explain why diamond has a very high melting point. *(3 marks)*

> Diamond has a giant covalent / macromolecular structure ✓ and so there are lots of strong covalent bonds holding the atoms together ✓, which requires lots of energy to break ✓, which is why diamond has a high melting point.

b) Explain why graphite is able to conduct electricity. *(2 marks)*

Delocalised electrons ✓ *are able to move through the structure and carry charge* ✓.

Graphite is unusual in that it has a giant covalent / macromolecular structure **and** it also conducts electricity.

c) Fullerene has a simple molecular structure. Explain why it has a lower melting point than either diamond or graphite. *(3 marks)*

Diamond and graphite both have giant covalent structures / covalent bonds holding the carbon atoms together ✓. In fullerenes there are intermolecular forces between the molecules ✓. Intermolecular forces are weaker than covalent bonds and so require less energy to overcome them ✓.

You are told in this question that fullerene has a simple molecular structure. Simple molecular structures have weak intermolecular forces.

These three questions are very similar to other questions earlier in this chapter. The point is that if you are able to recognise the type of structure present then you can apply your knowledge to the associated properties.

Example

Which one of the following statements is **incorrect**?

Tick **one** box. *(1 mark)*

Carbon nanotubes have a very high length to diameter ratio. ☐

Fullerenes are molecules of carbon atoms with hollow shapes. ☐

Fullerenes contain carbon atoms in a ring containing five, six or seven atoms. ☐

Graphene contains many layers of carbon atoms. ☐

Graphene contains many layers of carbon atoms. ✓

For more on the topics covered in this chapter, see pages 94–101 of the *Collins AQA GCSE Combined Science Revision Guide*.

10 Quantitative Chemistry

Chemists use quantities to determine the formula of compounds and when using chemical equations.

Chemical Measurements, Conservation of Mass and the Quantitative Interpretation of Chemical Equations

Chemical equations tell us about **mole** ratios and this idea is key to understanding this topic. For example, the equation

$$4Al + 3O_2 \rightarrow 2Al_2O_3$$

tells us that four moles of aluminium react with three moles of oxygen to form two moles of aluminium oxide.

Example

When magnesium (Mg) is heated in air it reacts with oxygen (O_2) to form magnesium oxide (MgO).

In an experiment, 0.45 g of magnesium was heated in air and the resulting magnesium oxide had a mass of 0.48 g.

a) Write a balanced chemical equation for the reaction of magnesium with oxygen. *(2 marks)*

$2Mg_{(s)} + O_{2(g)} \rightarrow 2MgO_{(s)}$
Correct symbols formulae ✓
Correctly balanced ✓

State symbols are not required in this answer but are included as an example of good practice.

b) Explain, using this example, what you understand by the 'law of conservation of mass'. *(2 marks)*

The law of **conservation of mass** states that no atoms are lost or made during a chemical reaction ✓, meaning that the mass of the **reactants** (magnesium and oxygen) must equal the mass of the product (magnesium oxide) ✓.

Because of the law of conservation of mass, we know that the total mass of products equals the total mass of the reactants. It follows from this (and the law of conservation of matter) that any atom that appears as a reactant must also appear as a product.

Quantitative Chemistry

c) Calculate the mass of oxygen that reacts with the magnesium in this experiment. *(1 mark)*

> $0.48\,g - 0.45\,g = 0.03\,g$ ✓

> The total mass of the products is $0.48\,g$ and therefore the total mass of the reactants must also equal $0.48\,g$.

Example

What is the **relative formula mass** of calcium nitrate $Ca(NO_3)_2$? Tick **one** box. *(1 mark)*

82 ☐

150 ☐

164 ☐

> 164 ✓

> Remember to use the mass number on the Periodic Table when doing any calculations that involve mass. There is one calcium atom, two nitrogen atoms and six oxygen atoms in the formula of calcium carbonate. Ensure you can work this out from the formula.

Example

In the equation below, the combined relative formula masses of all of the products is 157. Calculate the relative atomic mass of element X. *(2 marks)*

$$2HX + MgCO_3 \rightarrow MgX_2 + CO_2 + H_2O$$

> $157 - \{24 + 12 + (2 \times 16) + (2 \times 1) + 16\} = 71$ ✓
>
> $71/2 = 35.5$ ✓

> Take away the relative atomic masses of all of the other elements in the products from 157 and that will leave you the mass of the two atoms of X.
> Then divide by two to give you the relative atomic mass of X.

Example

In an experiment, a sample of copper (II) carbonate was placed in a crucible and heated for five minutes. In a separate experiment, a piece of copper metal was held in tongs and heated in a Bunsen flame for five minutes. The results of these experiments are shown in the table below.

Substance	Initial mass in g	Final mass in g
Copper (II) carbonate	6.20	4.00
Copper	6.35	7.95

a) Explain why the mass of copper (II) carbonate **decreased** during this experiment. *(2 marks)*

> The copper (II) carbonate thermally decomposes (breaks down due to the action of heat) ✓ to form copper (II) oxide and carbon dioxide. The carbon dioxide gas is released into the air and so the mass decreases ✓.

b) Write a word equation for the reaction occurring when copper (II) carbonate is heated. *(1 mark)*

> copper (II) carbonate → copper (II) oxide + carbon dioxide ✓

c) Explain how this experiment follows the law of conservation of mass. *(1 mark)*

> The mass of copper (II) carbonate equals the combined mass of the copper (II) oxide and carbon dioxide ✓.

d) Explain why the mass of copper **increased** during this experiment. *(2 marks)*

> Oxygen from the air ✓ reacts with / adds / bonds to the copper ✓.

e) Write a word equation for the reaction occurring when copper is heated. *(1 mark)*

> copper + oxygen → copper (II) oxide ✓ Copper oxide will be allowed.

Thermal decomposition of metal carbonates and the reaction of metals with oxygen are two examples of reactions that you need to be able to apply to the law of conservation of mass.

Copper can form two ions: Cu^+ and Cu^{2+}. When the Cu^{2+} ion is present, the name of the compound (such as the carbonate) will be written as copper (II) carbonate. This is to distinguish it from copper (I) carbonate, which contains the Cu^+ ion.

Use of Amount of Substance in Relation to Masses of Pure Substances

HT Many students find moles a hard topic. Moles are used in chemistry to represent the number of atoms, molecules or ions we have. The mass number on the Periodic Table tells you the mass of one mole of each element, e.g. 1 mole of magnesium has a mass of 24 g. A mole contains 6.02×10^{23} atoms so in the case of magnesium 24 g of magnesium means we have 6.02×10^{23} atoms.

HT Example

a) What is the numerical value of the **Avogadro constant** and what does this number represent? *(2 marks)*

6.02×10^{23} ✓. The number of atoms, molecules or ions in one mole of a given substance ✓.

Many GCSE pupils find 'moles' the hardest topic. This question is designed to help you get your head around what a mole is and how to use them. Essentially a mole is just a very big number!

b) Which one of the following contains the most atoms? Explain how you arrived at your answer. *(4 marks)*

In this question you need to convert mass into moles.

Moles = $\frac{mass}{M_r}$

M_r is the relative formula mass.

i) 12 g of C or 12 g of Mg?

Number of moles of carbon = $\frac{12}{12}$ = 1

Number of moles of magnesium = $\frac{12}{24}$ = 0.5

Correct working ✓

12 g of carbon contains more atoms because there are more moles of carbon than magnesium ✓.

ii) 16 g of sulfur or 13 g of aluminium?

Number of moles of sulfur = $\frac{16}{32}$ = 0.5

Number of moles of aluminium = $\frac{13}{27}$ = 0.48

Correct working ✓

16 g of sulfur contains more atoms because there are more moles of sulfur than aluminium ✓.

c) Which one of the following contains the most molecules? *(4 marks)*

i) 44 g of carbon dioxide (CO_2) or 9 g of water (H_2O)?

Number of moles of carbon dioxide = $\frac{44}{44}$ = 1

Number of moles of water = $\frac{9}{18}$ = 0.5 ✓

Therefore 44 g of carbon dioxide contains more molecules because there are more moles than water ✓.

ii) 13.75 g of phosphorus trichloride (PCl_3) or 23 g of nitrogen dioxide (NO_2)?

Number of moles of phosphorus trichloride = $\frac{13.75}{137.5}$ = 0.1

Number of moles of nitrogen dioxide = $\frac{23}{46}$ = 0.5 ✓

Therefore 23 g of nitrogen dioxide contains more molecules because there are more moles than phosphorus trichloride ✓.

> This question is effectively the same as the questions in part **b)**. Remember that moles can refer to atoms, molecules or ions.

d) Which one of the following contains the most atoms? *(4 marks)*

i) 44 g of carbon dioxide (CO_2) or 80 g of sulfur trioxide (SO_3)?

Number of moles of carbon dioxide = $\frac{44}{44}$ = 1

1 mole of CO_2 contains 3 moles of atoms (1 C and 2 O)

Number of moles of sulfur trioxide = $\frac{80}{80}$ = 1

1 mole of SO_3 contains 4 moles of atoms (1 S and 3 O) ✓

Therefore 80 g of SO_3 contains the most atoms ✓.

ii) 1 mole of calcium hydroxide ($Ca(OH)_2$) or 94 g of potassium oxide (K_2O)?

Number of moles of atoms in 1 mole of calcium hydroxide = 5 (1 Ca, 2 O and 2 H)

Number of moles of potassium oxide = $\frac{94}{94}$ = 1

1 mole of K_2O contains 3 moles of atoms (1 K and 2 O) ✓

Therefore 1 mole of calcium hydroxide contains the most atoms ✓.

> In part **d)** you need to calculate the number of moles and then look at how many atoms are present in each mole. One mole of carbon dioxide contains three moles of atoms (one mole of carbon and two moles of oxygen).

Quantitative Chemistry

HT **Example**

Consider the equation below.

$$Mg_{(s)} + 2HCl_{(aq)} \rightarrow MgCl_{2(aq)} + H_{2(g)}$$

Use the words in the box to complete the sentences about this equation. Each word can be used once, more than once or not at all.

one	two	three	four

a) In this equation, the number of moles of HCl that react with one mole of Mg is _____. *(1 mark)*

b) The total number of moles of reactants is _____. *(1 mark)*

c) The total number of moles of products is _____. *(1 mark)*

a) two ✓   **b)** three ✓   **c)** two ✓

This is a variation on a multiple-choice question. This question relies on you knowing that the number in front of each substance tells you the ratio of moles reacting. In this case one mole of magnesium (Mg) reacts with two moles of hydrochloric acid (HCl) to form one mole of magnesium chloride ($MgCl_2$) and one mole of hydrogen (H_2).

Reacting Mass Calculations

Moles of a substance can be calculated by dividing the mass of the substance by the mass of one mole of the substance, i.e.

$$\text{Moles} = \frac{\text{mass}}{M_r}$$

You need to be able to rearrange this equation.

 Example

What mass of calcium is required to produce 672 g of calcium oxide? *(3 marks)*

$$2Ca_{(s)} + O_{2(g)} \rightarrow 2CaO_{(s)}$$

Moles of calcium oxide =
$\frac{672}{56}$ = 12 ✓

Two moles of CaO are formed from two moles of Ca, i.e. 1:1 mole ratio. Therefore, moles of calcium required = 12 ✓

mass = moles × M_r

mass = 12 × 40 = 480 g ✓

This type of question is very common. Calculate the number of moles of the substance that you have the mass of, then deduce the number of moles of the substance you are asked about. Mass is calculated by multiplying the number of moles by the relative formula mass. The answer can also be found by using mass ratios.

	2Ca	O_2	2CaO
Reacting moles × M_r	2 × 40 = 80	(2 × 16) = 32	2 × (40 + 16) = 112
			× 6
Actual mass			672

Therefore, the mass of calcium required is 80 × 6 = 480 g

 Example

One of the main components of petrol is octane (C_8H_{18}). When petrol burns it reacts with oxygen in the air according to the equation below.

$$2C_8H_{18} + 25O_2 \rightarrow 16CO_2 + 18H_2O$$

A litre of petrol weighs 798 g. Calculate the mass of carbon dioxide produced when a litre of petrol is burnt. Assume that all of the petrol burns according to the above equation.

(3 marks)

Number of moles of petrol =
$\frac{798}{114} = 7$ ✓
(114 is the M_r of octane)
2 moles of petrol form 16 moles of CO_2, i.e. a 1:8 mole ratio.
Therefore, 7 moles of petrol will form 56 moles of CO_2 ✓
Mass of CO_2 = 56 × 44 = 2464 g / 2.464 kg ✓
(44 is the M_r of carbon dioxide)

Another example of this common question. Follow the steps given in the previous question. The ratio method is shown here as an alternative method.

	$2C_8H_{18}$	$25O_2$	$16CO_2$	$18H_2O$
Reacting moles × M_r	2 × [(8 × 12) + (18 × 1)] =	25 × (2 × 16) =	16 × [(12 + (2 × 16)] =	18 × [(2 × 1) + 16] =
	228	800	704	324
	× 3.5			
Actual mass	798			

Therefore, the mass of carbon dioxide formed will be 704 g × 3.5 = 2464 g / 2.464 kg.

Quantitative Chemistry

Metal X has a relative atomic mass of 23. X oxide has the formula X_2O. A student suggests that the chemical equation for the reaction between X and oxygen is

$$4X + O_2 \rightarrow 2X_2O$$

In an experiment 3.68 g of X was placed in a crucible and heated in air. The mass of metal oxide formed was 4.96 g.

a) What mass of oxygen reacted with metal X? *(1 mark)*

> $4.96\,g - 3.68\,g = 1.28\,g$

> This is another question applying the law of conservation of mass.

b) Use the results from the experiment to confirm that the above chemical equation is correct. *(3 marks)*

	X	O_2	X_2O	
Mass in g	3.68	1.28	4.96	
moles = $\frac{mass}{M_r}$	$\frac{3.68}{23}$ = 0.16	$\frac{1.28}{32}$ 0.04	$\frac{4.96}{62}$ 0.08	✓
Mole ratio	$\frac{0.16}{0.04}$ = 4	$\frac{0.04}{0.04}$ = 1	$\frac{0.08}{0.04}$ = 2	✓

> The mole ratio above is the same mole ratio as shown in the chemical equation ✓.

> A chemical equation tells you about reacting mole ratios, i.e. in the proposed chemical equation four moles of X react with one mole of O_2 forming two moles of X_2O. As you are given information about masses of chemicals reacting you need to convert this into moles. By dividing each number of moles by the smallest number (in this case 0.04) you obtain the mole ratio. The calculated mole ratio is the same as the mole ratio in the student's equation and therefore their suggested equation is correct.

10 g of hydrogen is added to 5 g of oxygen. The equation for the reaction is shown below.

$$2H_{2(g)} + O_{2(g)} \rightarrow 2H_2O_{(l)}$$

a) Identify the **limiting reactant**. *(3 marks)*

	H_2	O_2	
Mass in g	5	10	
moles = $\frac{mass}{M_r}$	$\frac{5}{2}$	$\frac{10}{32}$	
	= 2.5	= 0.3125	✓

The 2:1 mole ratio in the equation means that 2.5 moles of hydrogen requires 1.25 moles of oxygen to react ✓. Only 0.3125 moles of oxygen are present. As this is less than the required 1.25 moles, oxygen is the limiting reactant. ✓

The reactant that gets used up first in a reaction is the limiting reactant. You need to work in moles as the chemical equation tells you about reacting mole ratios.

b) Calculate the mass of water that will be produced in this reaction. *(3 marks)*

Oxygen is the limiting reactant. One mole of oxygen will form two moles of water. Therefore 0.3125 moles of oxygen will form 0.625 moles of water ✓. moles = $\frac{mass}{M_r}$, therefore moles × M_r = mass ✓ mass = 0.625 × 18 = 11.25 g ✓

The limiting reactant will determine the amount of products formed. A common error when calculating the mass of water formed is to multiply the number of moles by 36 instead of 18. Remember that mass = moles × M_r and M_r is the relative formula mass, i.e. the mass of **one** mole.

Moles in Solutions

The **concentration** of a solution in grams per dm^3 (g/dm^3) can be calculated using the following equation:

$$\text{concentration } (g/dm^3) = \frac{\text{mass (in g)}}{\text{volume (in } dm^3)}$$

Example

Calculate the mass in g of sodium hydroxide (NaOH) in 200 cm^3 of a 8g/dm^3 solution.

(2 marks)

mass = 8 × 0.2 ✓

= 1.6 g ✓

Remember that 1 dm^3 is 1000 cm^3. Therefore 200 cm^3 = 0.2 dm^3.

 For more on the topics covered in this chapter, see pages 102–105 of the *Collins AQA GCSE Combined Science Revision Guide*.

New substances are formed during chemical reactions. An understanding of how and why these reactions occur leads to the development of new materials and processes that will enhance our lives. For example, knowing how chemicals react based on their relative reactivities allows for metals to be **extracted** from ores obtained from the Earth.

Reactivity of Metals

The **reactivity series** is a list of metals placed in order of their reactivity. This can be used to help us to predict whether reactions will occur and to decide on the method of extracting metals from their ores.

Example

Small pieces of four different metals were placed in identical volumes of hydrochloric acid. The results are shown below.

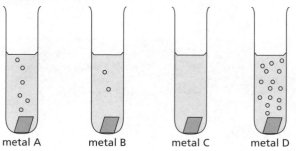

| metal A | metal B | metal C | metal D |

a) Place the metals in order of reactivity, starting with the most reactive. *(2 marks)*

Most reactive: D, A, B, C: Least reactive

All correct ✓✓

Two correct ✓

b) The four metals used in the experiment were calcium, magnesium, iron and copper.

Link the boxes connecting the letter of the metal to its identity. *(2 marks)*

metal A		calcium
metal B		magnesium
metal C		iron
metal D		copper

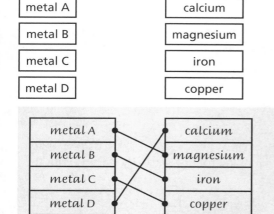

All correct ✓✓

Two correct ✓

a) and b) – you need to know the order of reactivity of the following eight metals (shown in order of decreasing reactivity):

potassium, sodium, lithium, calcium, magnesium, zinc, iron, copper.

c) Suggest why potassium was not used in this experiment. *(1 mark)*

Potassium reacts too violently with acid / too dangerous ✓.

d) Both calcium and magnesium form 2⁺ ions when they react. Which of these two metals has the greater tendency to form this ion? Explain your answer. *(2 marks)*

Calcium has a greater tendency to form a positive ion, as it is more reactive ✓.

More reactive metals have a greater tendency to form positive ions / it is easier for more reactive metals to lose their outer shell electrons ✓.

When metals react they form cations by losing outer shell electrons. The more reactive the metal, the more easily it will lose these electrons.

Chemical Changes

Example

A student investigated the different reactivities of a set of metals by placing pieces of each metal in metal nitrate solutions. Metal X is an unknown metal.

The table below shows some of the results.

Solution / metal	Copper	X	Iron	Magnesium
Copper (II) nitrate		✓	✓	
X (II) nitrate	✗		✗	✓
Iron (II) nitrate		✓		✓
Magnesium nitrate	✗	✗		

✓ = reaction observed   ✗ = no reaction

a) Use the results given to put the metals in order of reactivity starting with the most reactive.   *(2 marks)*

> Most reactive: magnesium, X, iron, copper: Least reactive
>
> All correct ✓✓
>
> Two correct ✓

Questions involving an understanding of displacement reactions are common. A displacement reaction occurs (shown by ✓) if the metal added is more reactive than the metal in solution.

b) Use the reactivity series in **a)** to complete the table.   *(3 marks)*

Solution \ metal	Copper	X	Iron	Magnesium
Copper (II) nitrate		✓	✓	✓
X (II) nitrate	✗		✗	✓
Iron (II) nitrate	✗	✓		✓
Magnesium nitrate	✗	✗	✗	

X is an unknown metal and so this question is beyond recall of the reactivity series. It is not displaced from its solution when magnesium is added meaning that X is less reactive than magnesium. X does displace iron from its solution so X is more reactive than iron. X is therefore between magnesium and iron in the reactivity series.

c) Write a word equation for the reaction of metal X with copper nitrate solution.   *(2 marks)*

> X + copper (II) nitrate ✓ → X (II) nitrate + copper ✓
>
> X + copper nitrate → X nitrate + copper will be allowed.

This is a word equation for the displacement reaction occurring.

d) Suggest another metal that will react with X nitrate solution. *(1 mark)*

Any metal above iron in the reactivity series (not including magnesium) ✓.

Example

Some metals, e.g. gold, are found in the Earth as the pure metal. Many metals such as zinc occur as oxides. Metals such as zinc can be extracted from their oxides by reaction with carbon. Some metals such as copper can be extracted from solutions by adding a more reactive metal such as iron.

a) State why metals such as gold are found as the pure metal. *(1 mark)*

They are unreactive / low in the reactivity series ✓.

Unreactive metals are found pure because they do not react with oxygen / water in the environment.

b) Write a word equation for the reaction of zinc oxide with carbon. *(2 marks)*

zinc oxide + carbon → zinc ✓ *+ carbon dioxide* ✓
Carbon monoxide will be allowed.

Carbon is often included in the reactivity series to provide a comparison. Metals below carbon in the reactivity series can be extracted from their compounds by **reduction** with carbon. Carbon is located above zinc in the reactivity series.

c) What substance has been reduced in this reaction? Explain your answer. *(2 marks)*

Zinc oxide ✓ *as it has lost oxygen* ✓.

Reduction involves the loss of oxygen.

d) **HT** When iron is added to copper (II) sulfate solution, iron (II) sulfate solution and copper metal are formed. Write an ionic equation, including state symbols, for this reaction. *(3 marks)*

$Fe_{(s)} + Cu^{2+}_{(aq)} → Fe^{2+}_{(aq)} + Cu_{(s)}$
Correct reactants ✓
Correct products ✓
Correct state symbols ✓

You need to be able to write ionic equations for displacement reactions. Write out all of the ions present and cancel any ions that appear on both sides of the equation (spectator ions), e.g. SO_4^{2-}.

e) **HT** In the answer equation for part **d)**, what has been oxidised? Explain your answer.

(2 marks)

> Fe has been oxidised ✓ because it loses electrons ✓.
>
> **Oxidation** is the loss of electrons. Fe has been oxidised because it turns into Fe^{2+} by losing two electrons.

Reactions of Acids

Salts are formed when acids are neutralised. You need to know the reactions of acids with metals, alkalis, bases and metal carbonates as well as how to prepare soluble salts. An accurate **neutralisation** can be carried out by **titration** and you need to be able to describe how to carry this out as well as do any associated calculations.

Example

What is the formula of the salt formed when magnesium reacts with sulfuric acid?
Tick **one** box. *(1 mark)*

MgS ☐ $MgSO_4$ ☐

MgS_2 ☐ $Mg(SO_4)_2$ ☐

> $MgSO_4$ ✓
>
> | Ion | Formula |
> | --- | --- |
> | Sulfate | SO_4^{2-} |
> | Nitrate | NO_3^- |
> | Carbonate | CO_3^{2-} |
> | Hydroxide | OH^- |
> | Ammonium | NH_4^+ |
>
> You need to be able to work out the ion formed by an element based on its location in the Periodic Table. You also need to know the ions listed in the table.

Example

Which combination of reactants would produce copper (II) sulfate and water as the only products? Tick **one** box. *(1 mark)*

Copper and sulfuric acid ☐

Copper (II) oxide and sulfuric acid ☐

Copper (II) carbonate and sulfuric acid ☐

Copper (II) hydroxide and nitric acid ☐

Copper (II) oxide and sulfuric acid ✓

Sulfuric acid forms sulfate salts (hydrochloric acid forms chlorides and nitric acid forms nitrates). To form a salt and water the reactants must be an acid with either a metal oxide or hydroxide. Copper (II) carbonate would react with sulfuric acid to form copper (II) sulfate and water but carbon dioxide would also be formed.

Example

Describe how a sample of pure copper (II) sulfate can be prepared in the laboratory starting from copper (II) oxide. Your answer should include a chemical equation and full experimental details. *(6 marks)*

$$CuO_{(s)} + H_2SO_{4(aq)} \rightarrow CuSO_{4(aq)} + H_2O_{(l)} ✓$$

State symbols are not required but are included for completeness.

Measure out sulfuric acid using a measuring cylinder ✓.

Transfer the acid to a beaker and warm the acid by gently heating over a Bunsen burner ✓.

Add spatula measures of copper (II) oxide powder and stir until no more dissolves ✓.

Filter the mixture ✓.

Use a water bath / electric heater to evaporate the solution and allow crystals to form ✓.

Soluble salts are always prepared by a chemical reaction followed by filtration and crystallisation.

As experimental details were asked for in this question, approach this by imagining that you are writing a method that someone else could follow.

Example

Which one of the following statements is **false**?

Tick **one** box. *(1 mark)*

All acids contain the H^+ ion. ☐

All alkalis contain the H^- ion. ☐

Aqueous solutions of acids have a pH of less than 7. ☐

The pH scale is from 0 to 14. ☐

All alkalis contain the H^- ion. ✓

Chemical Changes

Example

A titration was carried out to determine the concentration of a sample of sodium hydroxide solution. $25\,cm^3$ samples of the sodium hydroxide solution were titrated with hydrochloric acid of known concentration using phenolphthalein as an **indicator**. The results of the experiment are shown below.

	1	2	3	4
Initial burette reading / cm^3	0.00	24.20	0.00	23.50
Final burette reading / cm^3	24.20	47.10	23.50	46.30
Titre / cm^3	24.20			

a) Name the piece of apparatus used to record the volume of the hydrochloric acid solution. *(1 mark)*

> Burette ✓

> When an accurate but variable volume is required a burette is used.

b) Complete the table showing the titre values for titrations 2, 3 and 4. *(1 mark)*

> 2: 22.90   3: 23.50   4: 22.80
>
> All three correct ✓

> The numerical values are given to two decimal places and so you need to also give your titre values to two decimal places.

c) Titres should agree to within $0.20\,cm^3$. Calculate the mean titre for this titration. *(2 marks)*

> $\frac{22.90 + 22.80}{2}$ ✓ = $22.85\,cm^3$ ✓

> Only the second and fourth titration values should be used to calculate the mean titre. The first (rough) value should never be included and the third value is not a concordant value (i.e. it isn't within $0.20\,cm^3$ of the other results).

Example

Consider the information in the table below about two different acids A and B. One of the acids is citric acid and the other one is nitric acid.

Acid	A	B
Concentration (mol/dm³)	0.5	0.5
Colour of universal indicator in acid solution	Yellow	Red

a) Is acid A or acid B citric acid? Explain your answer. *(3 marks)*

> Acid A is citric acid ✓. The two acids have the same concentration but A has a higher pH (as shown by the colour of the universal indicator), meaning it must be the weak acid ✓. Citric acid is a **weak acid** / nitric acid is a **strong acid** ✓.

> The acids are the same concentration but clearly have different pH values. The less acidic one will be the weak acid.

b) What is meant by the term 'weak acid'? *(1 mark)*

> A weak acid partially **ionises** ✓ in aqueous solution.

> It is a common misconception that weak acids are those that are 'not very acidic', i.e. have a pH just under 7. The actual definition of weak acids refers to their degree of ionisation, which is how much the acids 'split up' into ions in solution.

HT **Example**

A pH probe inserted in 100 cm³ of a strong acid shows a reading of 4.00.

900 cm³ of water is added to the acid. What is the new reading on the pH probe? Tick **one** box. *(1 mark)*

3.00 ☐

4.00 ☐

4.40 ☐

5.00 ☐

> 5.00 ✓

> If the hydrogen ion concentration decreases by a factor of 10 then the pH increases by one unit.

Electrolysis

Electrolysis is the process of splitting up chemicals containing ions by passing electricity through them. This can be used to extract metals such as aluminium. Students often get confused because the term **cathode** refers to the negative **electrode** even though cations are positive. So remember that cations will be attracted to the cathode and anions will be attracted to the **anode**.

Example

An experiment was carried out to investigate the products of the electrolysis of molten lead (II) bromide ($PbBr_2$). The diagram below shows how the apparatus was set up.

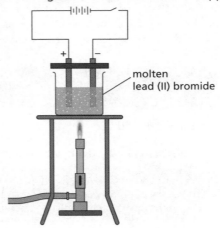

molten
lead (II) bromide

a) What name is given to the positive electrode? *(1 mark)*

Anode ✓

Confusion sometimes arises because positive ions are called cations but the positive electrode is called the anode! Remember 'PIANO – Positive Is Anode'.

b) Why does the lead (II) bromide need to be molten in order for electrolysis to occur? *(2 marks)*

So that the ions ✓ *are free to move* ✓.

A common error here is to answer the question in terms of electrons. When talking about the electrical conductivity of ionic compounds it is the ions that carry the charge. In a solid, the ions are unable to move and so ionic compounds can't carry the charge in the solid state.

c) Name the substance that will be formed at the positive electrode. *(1 mark)*

Bromine ✓

Negatively charged ions move to the positive electrode.

d) HT Write a **half equation** for the reaction occurring at the positive electrode. *(2 marks)*

$2Br^- \rightarrow Br_2 + 2e^-$ or $2Br^- - 2e^- \rightarrow Br_2$

Correct **species** in equation ✓

Balanced ✓

Ions gain or lose electrons when they reach the electrode and produce elements.

Example

The ore bauxite contains aluminium oxide. Aluminium is extracted from the ore by electrolysis. The diagram below represents the **cell** used to carry out this electrolysis.

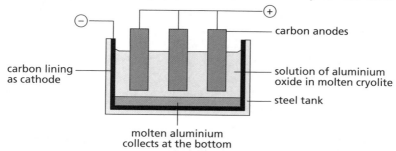

carbon anodes

carbon lining as cathode

solution of aluminium oxide in molten cryolite

steel tank

molten aluminium collects at the bottom

a) What is the formula of aluminium oxide? *(1 mark)*

Al_2O_3 ✓

b) Write a word equation for the overall reaction that occurs when aluminium oxide is electrolysed. *(1 mark)*

Aluminium oxide → aluminium + oxygen ✓

Electrolysis causes the ions to be discharged at the electrodes, producing elements.

c) State why aluminium is not extracted from bauxite by reduction with carbon. *(1 mark)*

Aluminium is more reactive / higher in the reactivity series ✓ (than carbon).

Carbon is unable to reduce metal oxides where the metal is higher than carbon in the reactivity series.

d) Why is the aluminium oxide dissolved in **cryolite**? *(2 marks)*

The mixture has a lower melting point than aluminium oxide alone ✓, which reduces the amount of energy required / lowers energy costs ✓.

Dissolving the aluminium oxide in cryolite does not lower the melting point of the aluminium oxide. It is the mixture that melts at a lower temperature than the aluminium oxide alone.

e) This extraction process uses large amounts of energy. Explain why so much energy is required in this process. *(2 marks)*

To produce the electrical current ✓ and to melt the (aluminium oxide / bauxite) mixture ✓.

f) Explain why the carbon anodes need to be regularly replaced. *(2 marks)*

The oxygen produced at the anode reacts with the carbon anode ✓ forming carbon dioxide gas, which leaves the cell ✓.

The oxygen produced at the anode reacts exothermically (i.e. produces heat) with the carbon anodes. This cheap source of heat is used to help keep the mixture molten.

g) **HT** Write a half equation for the formation of oxygen at the anode. *(2 marks)*

$2O^{2-} \rightarrow O_2 + 4e^-$ or $2O^{2-} - 4e^- \rightarrow O_2$

Correct species in equation ✓

Balanced ✓

h) HT Is the reaction in part **g)** a reduction or oxidation reaction?
  Explain your answer.   *(2 marks)*

Oxidation ✓ because the oxide ions lose electrons ✓.

OILRIG – Oxidation Is Loss (of electrons) Reduction Is Gain (of electrons).

Example

What are the products formed at the anode and cathode when potassium fluoride
solution is electrolysed?   *(1 mark)*

	Anode	Cathode
A	Oxygen	Hydrogen
B	Fluorine	Hydrogen
C	Oxygen	Potassium
D	Fluorine	Potassium

B Fluorine   Hydrogen ✓

During electrolysis of aqueous solutions at the cathode the least reactive element is formed
and at the anode oxygen is formed unless the solution contains **halide** ions.

Example

An experiment was carried out to identify the products of the electrolysis of aqueous
sodium chloride. The apparatus below was assembled.

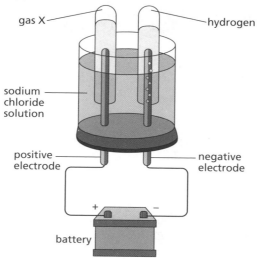

gas X ⎯ hydrogen

sodium
chloride
solution

positive
electrode

negative
electrode

battery

Chemical Changes

a) Identify gas X. *(1 mark)*

Chlorine ✓

A halide ion is present and so the halogen will be formed at the anode.

b) HT Write a half equation for the formation of hydrogen. *(2 marks)*

$2H^+ + 2e^- \rightarrow H_2$

Correct species ✓

Balanced ✓

Hydrogen gas is diatomic so two H^+ ions are required.

c) Name the solution formed during this electrolysis. *(1 mark)*

Sodium hydroxide ✓

Sodium ions and hydroxide ions are not discharged during this electrolysis and so they remain in solution.

d) What colour will universal indicator turn when added to the remaining solution? Explain your answer. *(2 marks)*

Dark green / blue / purple ✓ because the hydroxide / OH^- ion remains / the solution becomes alkaline ✓.

The presence of hydroxide means that the solution will be alkaline.

 For more on the topics covered in this chapter, see pages 114–119 of the *Collins AQA GCSE Combined Science Revision Guide*.

Energy is usually given out (**exothermic**) or taken in (**endothermic**) during chemical reactions. These transfers of energy are a result of the difference in energy between bonds being broken in reactant molecules and bonds being made in product molecules.

Exothermic and Endothermic Reactions

Exothermic reactions are those that transfer energy into the surroundings and so the temperature of the surroundings increases. Endothermic reactions are the opposite, absorbing energy from the surroundings and therefore the temperature of the surroundings decreases.

Example

Which one of the following statements is **incorrect**? Tick **one** box.   *(1 mark)*

After an exothermic reaction has taken place there is more energy in the universe.

If more energy is required to break the bonds in the reactant molecules than is released when bonds in the product molecules are made, then the reaction is endothermic.

During an exothermic reaction the temperature of the surroundings increases.

In an endothermic reaction the products have more energy than the reactants.

After an exothermic reaction has taken place there is more energy in the universe. ✓

Energy is conserved in chemical reactions. The amount of energy transferred to the surroundings is the same as the amount of energy lost by the reaction.

Example

Which one of the following reactions is an example of an endothermic chemical reaction? Tick **one** box.   *(1 mark)*

Combustion

Neutralisation

The reaction between citric acid and sodium hydrogencarbonate

Melting ice

Energy Changes

The reaction between citric acid and sodium hydrogencarbonate ✓

Melting ice does require energy **but** it is a physical change and not a chemical reaction.

Example

An experiment was carried out to investigate the changes of temperature that occur during a neutralisation reaction. The following method was carried out.

Step 1: 30 cm³ of hydrochloric acid was measured out.

Step 2: The hydrochloric acid was transferred to a polystyrene cup.

Step 3: A thermometer was placed in the acid and the temperature recorded.

Step 4: 5 cm³ of sodium hydroxide was added to the hydrochloric acid.

Step 5: The mixture was stirred and when the temperature stopped changing it was recorded.

Step 6: Steps 4 and 5 were repeated until 50 cm³ of sodium hydroxide had been added.

Step 7: The experiment was repeated.

The results of the experiment are shown below.

Total volume of sodium hydroxide added in cm³	Maximum temperature in °C		
	First trial	**Second trial**	**Mean**
0	20	22	
5	25	25	
10	29	31	
15	33	35	
20	39	39	
25	41	43	
30	46	46	
35	46	44	
40	34	36	
45	27	25	
50	17	17	

a) In step 1 suggest a suitable piece of apparatus to measure out the hydrochloric acid. *(1 mark)*

> Pipette / burette / measuring cylinder ✓

b) Explain why a polystyrene cup was used in step 2. *(2 marks)*

> It is an insulator ✓ and so keeps the heat in / reduces heat loss ✓.

c) Why was the mixture stirred in step 5? *(1 mark)*

> To ensure that the reaction occurs / to ensure complete reaction / to distribute the heat evenly ✓.

d) Complete the column in the table showing the mean temperature. *(1 mark)*

> 21, 25, 30, 34, 39, 42, 46, 45, 35, 26, 17 ✓

e) **i)** Plot a graph of the 'mean temperature' against 'total volume of sodium hydroxide added' on the axes below. *(3 marks)*

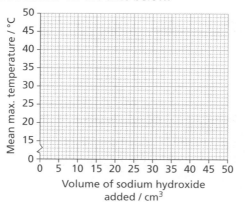

ii) Draw **two** straight lines of best fit:

- one through the points that are increasing
- one through the points that are decreasing.

Ensure the two lines are extended so they cross each other. *(2 marks)*

Energy Changes

i)

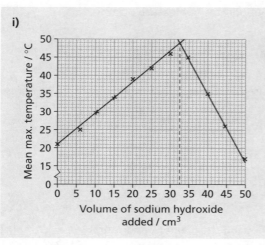

✓ ✓ ✓ for all points plotted correctly.
✓ ✓ for one point plotted incorrectly.
✓ for two points plotted incorrectly.

ii) Correct line of best fit for increasing values ✓

Correct line of best fit for decreasing values ✓

f) Use your graph to estimate the volume of sodium hydroxide required to neutralise the hydrochloric acid. Show your working on the graph. *(2 marks)*

> Working ✓ 33 cm³ ✓
>
> Range 32–34 cm³ will be allowed.

g) Explain why the temperature decreases after 35 cm³ of sodium hydroxide was added. *(2 marks)*

> *The reaction is complete / no further reaction takes place ✓ and the temperature decreases because a cooler solution is being added to a warmer solution ✓.*

This question focuses on one of the required practical tasks. You will have carried out a practical very similar to this and so questions could be asked that directly relate to carrying out the practical and the analysis and interpretation of results.

Marks are often lost when drawing graphs because the points are not plotted accurately. Take your time when doing this.

Example

Consider the **reaction profile** shown.

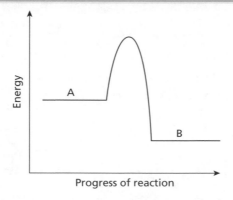

a) On the diagram what is represented by the letters A and B? *(1 mark)*

A = reactants, B = products ✓

b) Does the above reaction profile represent an exothermic or endothermic reaction? Explain your answer. *(2 marks)*

Exothermic ✓ because the products are lower in energy than the reactants ✓.

c) On the diagram show and label:

  i) the **activation energy** for the reaction. *(1 mark)*

  ii) the overall energy change (ΔH). *(1 mark)*

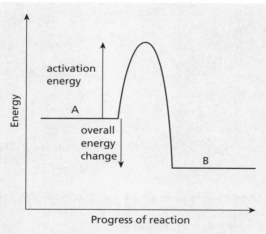

i) Correctly drawn / labelled arrow for activation energy ✓

ii) Correctly drawn / labelled arrow for overall energy change ✓

The minimum amount of energy that particles must have to react is called the activation energy. This is the amount of energy needed to break the bonds in the reactant molecules and it is the energy increase between the reactants and the top of the reaction profile.

HT **Example**

The equation for the combustion of propane is written below, showing the structural formulae of the reactants and products.

$$H-\underset{\underset{H}{|}}{\overset{\overset{H}{|}}{C}}-\underset{\underset{H}{|}}{\overset{\overset{H}{|}}{C}}-\underset{\underset{H}{|}}{\overset{\overset{H}{|}}{C}}-H \;+\; 5\,O{=}O \;\longrightarrow\; 3\,O{=}C{=}O \;+\; 4\,H-O-H$$

Energy Changes

a) Use the bond energies given in the table to help you to calculate the energy change for this reaction. *(3 marks)*

Bond	Bond energy in kJ
C-H	413
C-C	347
O=O	498
C=O	805
H-O	464

Bonds broken:

2 C–C   2 × 347 kJ

8 C–H   8 × 413 kJ

5 O=O   5 × 498 kJ

Total: 6488 kJ ✓

Bonds formed:

6 C=O   6 × 805 kJ

8 O–H   8 × 464 kJ

Total: 8542 kJ ✓

The overall energy change (ΔH) is calculated by bonds broken – bonds formed.

Energy change = –2054 kJ ✓

b) What is the numerical value of the activation energy for this reaction? *(1 mark)*

6488 kJ / value for the 'bonds broken' from part a) ✓

c) In terms of the bond energies explain why this reaction is exothermic. *(1 mark)*

More energy is released when the bonds in the product molecules are made than is used to break the bonds in the reactant molecules ✓.

When carrying out bond energy calculations it is a good idea to cross off the bonds in the displayed structural formulae of the molecules. That way you are unlikely to accidentally miss bonds. Remember that the numbers in the equation show the number of molecules present.

For more on the topics covered in this chapter, see pages 120–123 of the *Collins AQA GCSE Combined Science Revision Guide*.

13 The Rate and Extent of Chemical Change

Rates of reaction can be explained by collision theory. This is a practical-based topic and so questions based on experiments are commonly asked in exams.

Rate of Reaction

The rate of reaction is a measure of how much reaction is used up or product formed in a given amount of time. The rate of a particular reaction can be affected by temperature, concentration, pressure (if the reactants are gases), a **catalyst** and surface area.

Example

A student carried out an experiment investigating the rate of the reaction between magnesium and hydrochloric acid.

A diagram of the apparatus used is shown below.

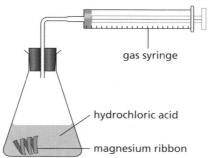

gas syringe

hydrochloric acid

magnesium ribbon

The student recorded the volume of gas collected every 10 seconds. A graph of the results is shown below.

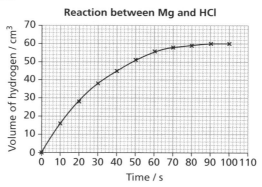

Reaction between Mg and HCl

The Rate and Extent of Chemical Change

a) Name the gas produced in this experiment. *(1 mark)*

> Hydrogen ✓

b) What volume of this gas was collected in the first minute of the experiment? *(1 mark)*

> 57 cm³ ✓
> Range 56–58 cm³ will be allowed.

> It is a good idea to show your working by drawing on the graph. Use a ruler to draw a line from 60 seconds up to the curve and then across to the volume.

c) After 20 seconds the volume of gas collected was 28 cm³. Calculate the mean rate of reaction during the first 20 seconds. *(2 marks)*

> 28 ÷ 20 = 1.4 ✓ cm³/s ✓
> cm³s⁻¹ will be allowed.

> Remember the units!

d) In the next 20 seconds the mean rate of reaction was 0.95 cm³/s. Explain why the rate of reaction was less than during the first 20 seconds. *(2 marks)*

> The concentration of acid had decreased ✓ and so there were fewer collisions between the magnesium and acid ✓.

> Any questions asking you to explain rates of reaction should include the idea of 'collisions'.

e) **HT** By plotting a **tangent** on the graph, calculate the rate of reaction after 40 seconds. *(3 marks)*

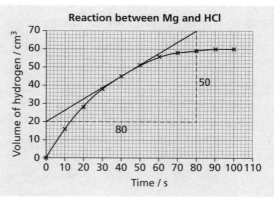

Draw a tangent to the curve at 40 seconds ✓.

Draw a triangle using your tangent as the hypotenuse ✓.

Determine the change in y and the change in x;

determine the gradient by calculating:

$\frac{\text{change in y}}{\text{change in x}}$ ✓

Any answer that is derived from a correctly drawn tangent and correct subsequent calculations will be allowed, e.g.

change in y = 50 cm³

change in x = 80 s

rate = $\frac{50}{80}$

= 0.63 cm³/s

Make sure you know how to calculate the **gradient** from a tangent.

f) How long did it take for the reaction to complete? Explain your answer.   *(2 marks)*

90 seconds ✓ No more gas is being produced after this time / the graph is horizontal / one of the reactants has been used up ✓.

Again – show your working by drawing on the graph. Decide where you think the graph becomes flat (as this occurs when the reaction has finished) and draw a line down to the x-axis to determine the time.

Example

An experiment was carried out to investigate the effect of changing the surface area on the rate of reaction between hydrochloric acid and calcium carbonate.

The equation for the reaction is:

$$CaCO_{3(s)} + 2HCl_{(aq)} \rightarrow CaCl_{2(aq)} + H_2O_{(l)} + CO_{2(g)}$$

25 cm³ of hydrochloric acid at a temperature of 20°C was put into a conical flask and placed on a balance. 5 g of calcium carbonate chips were added to the flask and a piece of cotton wool loosely placed in the neck of the flask.

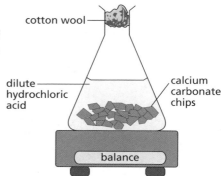

cotton wool

dilute hydrochloric acid

calcium carbonate chips

balance

A diagram of the apparatus set up is shown here.

The Rate and Extent of Chemical Change

The initial mass was recorded, a timer started and the mass recorded every 20 seconds for two minutes. At the end of the experiment a small amount of calcium carbonate remained. The results are shown in the table below.

Time/s	0	20	40	60	80	100	120
Mass/g	101.60	101.40	101.27	101.18	101.12	101.10	101.10

This is a very common experiment when looking at rates of reaction. The rate of reaction is measured by considering how quickly the carbon dioxide gas is produced.

a) Explain the purpose of the cotton wool. *(2 marks)*

To allow the gas to escape ✓ but to keep the liquid inside / prevent the liquid spitting out ✓.

The gas escapes through the cotton wool into the atmosphere but any liquid splashes are kept inside the apparatus. If any liquid escapes, it would give an artificially high mass loss.

b) State why the mass decreases during the experiment. *(1 mark)*

The gas leaves the apparatus / escapes into the atmosphere ✓.

c) Plot a graph of the results on the axes below. Connect your points using a smooth curve. *(3 marks)*

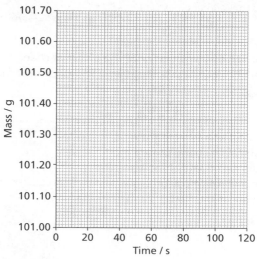

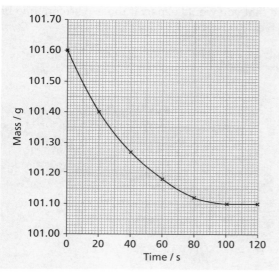

Correctly plotted points ✓✓

Smooth curve through all / close to the points ✓

d) How can you tell that the reaction has finished? Explain your answer.   *(1 mark)*

The mass no longer decreased / changed / the graph eventually flattens out / remains constant ✓.

When no more gas is produced the mass remains constant meaning that the reaction has finished.

e) The rate of reaction is higher during the first 20 seconds than during the second 20 seconds.

i) How does your graph show this?   *(2 marks)*

The curve is steeper ✓ during the first 20 seconds than the second 20 seconds ✓.

The steeper the slope of the curve then the greater the rate of reaction.

ii) Explain why the rate of reaction decreases as the reaction proceeds.   *(2 marks)*

The concentration of the (hydrochloric) acid decreases ✓ and so there are fewer (successful) collisions with the calcium carbonate particles ✓.

The calcium carbonate is in excess (you know this because there is calcium carbonate left over at the end of the experiment) and so it must be the acid that is used up.

f) Draw on your graph the curve that you would expect if the experiment had been carried out at 40°C rather than at 20°C. Label the curve '40°C'. *(2 marks)*

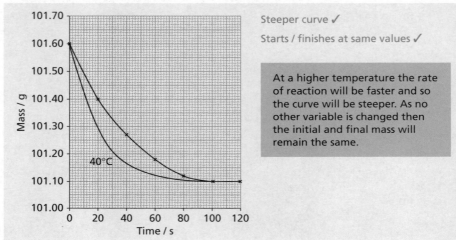

Steeper curve ✓

Starts / finishes at same values ✓

At a higher temperature the rate of reaction will be faster and so the curve will be steeper. As no other variable is changed then the initial and final mass will remain the same.

Example

Sodium thiosulfate solution reacts with hydrochloric acid. As the reaction takes place the solution slowly turns cloudy.

In an experiment 5 cm³ of dilute hydrochloric acid was added to 50 cm³ of sodium thiosulfate solution in a conical flask. The flask was placed on a piece of paper with an 'X' written on it in black pen. The student looked through the conical flask at the 'X' and timed how long it took for it to no longer be visible.

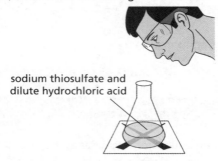

sodium thiosulfate and dilute hydrochloric acid

The experiment was repeated with sodium thiosulfate solution of different concentrations.

The equation for the reaction is:

$$Na_2S_2O_{3(aq)} + 2HCl_{(aq)} \rightarrow S_{(s)} + SO_{2(g)} + 2NaCl_{(aq)} + H_2O_{(l)}$$

A graph of the results is shown below.

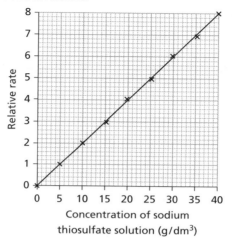

Concentration of sodium
thiosulfate solution (g/dm³)

a) Explain why the solution turns cloudy. *(2 marks)*

> Solid / a **precipitate** ✓ sulfur ✓ is produced (which prevents the light
> passing through).

Looking at the equation provided, and particularly the state symbols, will often help you
when answering questions that ask you about what you would see during a reaction.

b) The student concluded that 'the rate of reaction is directly proportional to the
concentration of the sodium thiosulfate solution'. How does the graph support
this conclusion? *(1 mark)*

> As the concentration doubles so does the (relative) rate ✓.

If two quantities are in direct proportion, as one increases, the other increases by the
same percentage.

c) What is the relative rate of reaction when the concentration of sodium
thiosulfate is 12 g/dm³? *(1 mark)*

> 2.40 ✓
> 2.30–2.50 is acceptable.

Draw a line on the graph from 12 g/dm³ on the x-axis up to the line and then across to
the y-axis to determine the relative rate.

d) Explain why the rate of reaction increases when the concentration of sodium thiosulfate solution increases. *(2 marks)*

> There is a greater frequency ✓ of collisions (between reacting particles) ✓.

Always try and answer questions explaining rates of reaction by talking about 'collisions'.

Example

A student makes the following statement:

> 'When the temperature increases the rate of reaction increases because there is a higher frequency of collisions.'

To what extent do you support this statement? *(3 marks)*

> Increasing temperature does lead to a higher frequency of collisions / increased rate of reaction ✓.
>
> The statement is partially correct / doesn't fully explain the effect of temperature on rate of reaction ✓.
>
> At higher temperatures, there are more energetic collisions / more particles possess energy equal to or greater than the activation energy ✓.

Temperature not only increases the *frequency* of collisions (because particles have more kinetic energy) but it also increases the number of *successful* collisions (because more particles have energy equal to or greater than the energy required to react).

Example

Hydrogen peroxide, H_2O_2, decomposes very slowly into water and oxygen. The catalyst manganese dioxide, MnO_2, is frequently added to speed up the decomposition process.

The equation for the decomposition of hydrogen peroxide is:

$$2H_2O_{2(aq)} \rightarrow 2H_2O_{(l)} + O_{2(g)}$$

The reaction profile for the reaction is shown below.

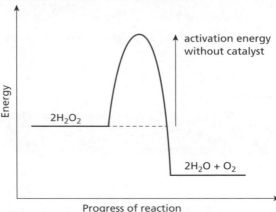

a) What is a catalyst? *(2 marks)*

A species / substance that changes / speeds up the rate of chemical reactions ✓ but is not used up during the reaction ✓.

b) Explain how a catalyst works. *(2 marks)*

Catalysts provide a different / alternative pathway ✓ for the reaction that has a lower activation energy ✓.

For **a)** and **b)** make sure you can distinguish between what a catalyst is and how a catalyst works. How it works is asked for in part **b)**.

c) Draw on the diagram the reaction profile when the manganese dioxide catalyst is present. *(2 marks)*

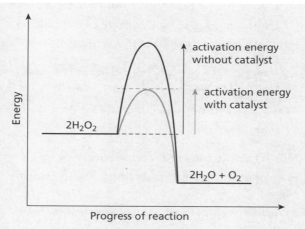

Start and finish on the reactants / products line ✓

Curve is smaller than the original / lower activation energy ✓

The idea that the catalyst lowers the activation energy for the reaction is shown on the reaction profile.

d) How can you tell from the equation for the reaction that manganese dioxide is a catalyst and not a reactant? *(1 mark)*

The catalyst is not included in the chemical equation for the reaction ✓.

As catalysts are not used up during the reaction they do not appear in chemical reactions. They can be shown on the arrow between the reactants and products.

e) A student carried out an investigation into the effect of particle size of manganese dioxide on the decomposition of hydrogen peroxide. Her results table is shown below.

Experiment	Method	Observations
1	I added a 1.0g piece of manganese dioxide to 10cm³ of 34g/dm³ hydrogen peroxide solution	There was slow fizzing and bubbling
2	I added 1.0g of manganese dioxide powder to 10cm³ of 34g/dm³ hydrogen peroxide solution	There was rapid fizzing and bubbling

i) State two ways that the student ensured a fair test. *(2 marks)*

✓✓ Two from:

Used the same mass of catalyst / manganese dioxide.

Used the same volume of hydrogen peroxide solution.

Used the same concentration of hydrogen peroxide solution.

In an experiment where one variable is changed (in this case the surface area of the catalyst) all other factors that could affect the rate of the reaction must be kept constant.

ii) Explain her observations. *(4 marks)*

Bubbling is due to the production of oxygen / gas ✓.

The piece of manganese dioxide has a smaller surface area than the powder ✓.

Therefore, the piece of manganese dioxide has fewer sites available for collisions to occur ✓.

Meaning there is a slower rate of reaction ✓.

In this question you were asked to explain the observations – that should include why there were bubbles as well as explaining the different rate of their production.

Reversible Reactions and Dynamic Equilibrium

In a **reversible reaction** the reactants react to form products but also the products can react to re-form the reactants. When the rate of the forward reaction is the same as the rate of the reverse reaction then a state of dynamic **equilibrium** exists. Changing the temperature, concentration or pressure of a reaction at equilibrium can change the amount of product formed in the reaction.

Example

Which one of the following statements about reversible reactions is **false**?
Tick **one** box.   *(1 mark)*

The direction of reversible reactions can be changed by changing
the conditions. ☐

If the forward reaction is endothermic then the reverse reaction is exothermic. ☐

A reversible reaction is at equilibrium when there are equal amounts of
reactants and products. ☐

For equilibrium to occur then the rate of the forward reaction must be
the same as the rate of the reverse reaction. ☐

> A reversible reaction is at equilibrium when there are equal amounts of
> reactants and products. ✓

Example

Solid ammonium chloride decomposes upon heating to form ammonia gas and hydrogen chloride gas. These two gases combine and re-form ammonium chloride.

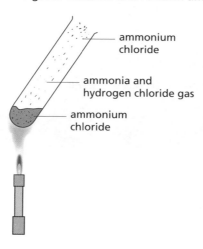

ammonium chloride

ammonia and hydrogen chloride gas

ammonium chloride

The Rate and Extent of Chemical Change

a) Write a word equation for this reaction. *(2 marks)*

ammonium chloride ⇌ ammonia + hydrogen chloride

Correct names on the correct side of the equation ✓

Correct symbol, i.e. ⇌ ✓

Because the reaction can go forwards and backwards it is a reversible reaction.

b) Is the reaction between ammonia and hydrogen chloride exothermic or endothermic? Explain your answer. *(2 marks)*

Exothermic ✓. If the forward reaction is endothermic (i.e. requires heat) then the reverse reaction is exothermic ✓.

In a reversible reaction if one reaction is endothermic then the other reaction is exothermic. The same amount of energy is transferred in each case.

Example

When **hydrated** copper sulfate is heated, water is given off and anhydrous copper sulfate remains. This is a reversible reaction.

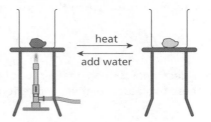

a) Write a word equation for this reaction. *(2 marks)*

hydrated copper sulfate ⇌ anhydrous copper sulfate + water

Correct names on the correct side of the equation ✓

Correct symbol, i.e. ⇌ ✓

This reaction is one that you are specifically required to know.

b) Anhydrous copper sulfate was placed in a test tube and 20 drops of water added. The mixture was carefully stirred with a thermometer. Describe what observations would be made during this experiment. *(2 marks)*

The white anhydrous copper sulfate would turn blue ✓.

The temperature would rise / increase ✓.

You also need to know the colours of anhydrous / hydrated copper sulfate. A thermometer was used to stir the mixture meaning that you would expect a temperature change! If the forward reaction is endothermic then the reverse reaction is exothermic; hence a temperature increase would be observed.

 Example

Which one of the following statements about equilibrium is correct?
Tick **one** box. *(1 mark)*

Reactants and products may leave the apparatus where the reversible reaction is occurring. ☐

The relative amounts of all the reactants and products at equilibrium do not depend on the conditions of the reaction. ☐

If the concentration of a reactant is increased then the concentration of the other reactants will also increase. ☐

Le Chatelier's Principle states that if a system is at equilibrium and a change is made to any conditions then the system responds to counteract the change. ☐

Le Chatelier's Principle states that if a system is at equilibrium and a change is made to any conditions then the system responds to counteract the change. ✓

 Example

Consider the following reversible reaction.

$$2NO_{2(g)} \rightleftharpoons N_2O_{4(g)} \quad \Delta H = -58\,kJ/mol$$

brown   colourless

At 20°C a sealed tube containing a mixture of both gases at equilibrium is pale brown in colour.

a) What is meant by the term 'equilibrium'? *(2 marks)*

A reversible reaction ✓ where the forward and reverse reactions occur at exactly the same rate ✓.

b) What will be observed if the equilibrium mixture is placed into ice cold water? Explain your answer with reference to Le Chatelier's Principle. *(3 marks)*

The colour will become paler / less brown / colourless ✓.

Decreasing temperature favours the exothermic reaction ✓, which is the forward reaction, meaning that the relative amounts of products increases, which will cause the mixture to become lighter in colour ✓.

c) How can the pressure be adjusted to make the mixture go darker in colour? Explain your answer. *(2 marks)*

Decrease / lower the pressure ✓.

Decreasing the pressure causes the equilibrium position to shift towards the side with the larger number of molecules as shown by the symbol equation for that reaction ✓.

Questions on equilibrium are always challenging. Consider the following key points when answering questions on equilibrium:

- increasing the concentration of a reactant means that more products will be formed until equilibrium is reached again
- increasing the temperature will favour the endothermic reaction
- increasing the pressure will favour the reaction that produces fewer molecules of gas (as shown by the symbol equation).

 Example

When carbon dioxide gas dissolves in water the following equilibria are established:

$$CO_{2(g)} \rightleftharpoons CO_{2(aq)}$$

$$CO_{2(aq)} + H_2O_{(l)} \rightleftharpoons H_2CO_{3(aq)}$$

Predict the effect on the concentration of $H_2CO_{3(aq)}$ when the concentration of $CO_{2(g)}$ is increased. Explain your answer. *(3 marks)*

The concentration of $H_2CO_{3(aq)}$ will increase ✓.

Increasing the concentration of $CO_{2(g)}$ in the first equation will result in an increase of $CO_{2(aq)}$ ✓.

$CO_{2(aq)}$ is a reactant in the second equation and so an increase in the concentration of $CO_{2(aq)}$ will result in an increase in the concentration of $H_2CO_{3(aq)}$ ✓.

In this question, the two equilibria are related because $CO_{2(aq)}$ is a product in the first equation and then a reactant in the second equation. Remember that increasing the concentration of a reactant means that more products will be formed until equilibrium is reached again.

HT Example

Consider the equilibrium shown below.

$$CH_{4(g)} + H_2O_{(g)} \rightleftharpoons CO_{(g)} + 3H_{2(g)} \quad \Delta H = 206\,kJ/mol$$

Which set of conditions will maximise the formation of the products?
Tick **one** box.   *(1 mark)*

High temperature and high pressure ☐

High temperature and low pressure ☐

Low temperature and high pressure ☐

Low temperature and low pressure ☐

High temperature and low pressure ✓

The forward reaction is endothermic (ΔH has a positive value) and high temperatures favour endothermic reactions. There are two molecules of gas on the left-hand side of the equation and four on the right-hand side. A low pressure will favour the reaction that produces more molecules of gas.

HT Example

Ammonia, NH_3, is produced by reacting together nitrogen and hydrogen. The equation for the reaction is shown below.

$$N_{2(g)} + 3H_{2(g)} \rightleftharpoons 2NH_{3(g)}$$

The graph below shows the yield (amount) of ammonia produced at varying temperatures and pressures.

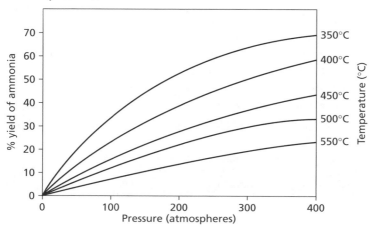

The Rate and Extent of Chemical Change

Use the graph to help you answer the following questions.

a) What conditions are required to obtain the highest yield of ammonia? *(2 marks)*

Temperature of 350°C ✓. Pressure of 400 atmospheres ✓.

b) What is the effect of increasing pressure on the yield of ammonia?
Explain why this effect occurs. *(3 marks)*

Increasing pressure increases the yield of ammonia ✓. Increasing pressure favours the reaction that produces fewer molecules of gas ✓. There are four molecules of gas on the left-hand side of the equation and two on the right ✓.

c) Is the formation of ammonia an exothermic or endothermic process?
Explain your answer. *(3 marks)*

Exothermic ✓ because at a fixed pressure a higher temperature produces a lower yield of ammonia ✓ and increasing the temperature favours the endothermic reaction ✓.

The graph at first appears complicated because it is showing three variables – yield of ammonia, pressure and temperature. To look at the effect of temperature on yield choose a particular pressure, and to look at the effect of pressure on yield choose a fixed temperature.

HT Example

Consider the equilibrium below.

$$2SO_{2(g)} + O_{2(g)} \rightleftharpoons 2SO_{3(g)} \quad \Delta H = -196 \text{ kJ/mol}$$

Explain why it is difficult to predict the effect on the yield of SO_3 when both the pressure and temperature are increased. *(3 marks)*

Increasing the pressure alone will increase the yield of SO_3 ✓ (because there are less molecules of gas on the right-hand side of the equation).

Increasing the temperature alone will decrease the yield of SO_3 ✓ (because increasing the temperature favours the endothermic reaction).

The two changes have opposite effects and you do not know which change will have the greater effect ✓.

Break this question down by thinking about the two changes one at a time. That should lead you to the idea that the two changes have opposing effects.

For more on the topics covered in this chapter, see pages 124–127 of the *Collins AQA GCSE Combined Science Revision Guide*.

14 Organic Chemistry

The chemistry of carbon compounds is one of the most important areas of chemistry. Many everyday materials including plastics and medicines are made from organic molecules. The main feedstock for these materials is crude oil, which is a finite resource.

Carbon Compounds as Fuels and Feedstock

Crude oil is a mixture of **hydrocarbon** molecules that can be separated out by fractional distillation into different fractions. Some of these are used as fuels and others are used to make other chemicals.

Example

Which one of the following does **not** represent an **alkane**?
Tick **one** box. *(1 mark)*

C_nH_{2n+2} ☐

CH_4 ☐

$$H-\overset{\displaystyle H}{\underset{\displaystyle H}{C}}-\overset{\displaystyle H}{\underset{\displaystyle H}{C}}-H$$ ☐

C_3H_6 ☐

C_3H_6 ✓

Example

Which of the following is **not** a name of one of the fractions obtained from crude oil?
Tick **one** box. *(1 mark)*

Kerosene ☐

Petrol ☐

Hydrocarbons ☐

Heavy fuel oil ☐

Hydrocarbons ✓

Organic Chemistry

Example

Which is the third member of the alkanes? Tick **one** box.   *(1 mark)*

Butane ☐

Ethane ☐

Methane ☐

Propane ☐

> *Propane* ✓
>
> These three multiple choice questions test your recall of facts about crude oil and alkanes.
> Hydrocarbons are obtained from crude oil but they are not a specific fraction of their own.
> Most of the molecules in all of the fractions are hydrocarbons called alkanes.

Example

Crude oil can be separated by a fractionating column in a process known as fractional distillation.

The diagram below shows a simplified fractionating column.

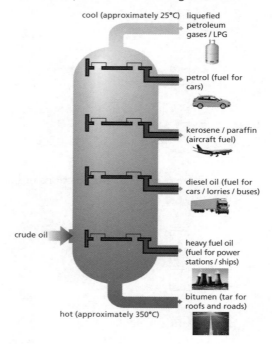

a) Explain how crude oil is separated into its individual fractions. *(4 marks)*

> The crude oil is heated until it boils / evaporates / turns into a vapour ✓.
> It then enters the fractionating column where there is a temperature gradient ✓.
> The vapours rise up the column until the temperature equals their boiling point ✓.
> The vapours condense at their boiling point ✓ and are collected / separated out.

You need to know the details of how fractional distillation works. Think of the different stages that happen to the molecules. The key idea in fractional distillation is that molecules are separated out according to their boiling points.

b) Explain how the following are related to the size of hydrocarbon molecules. *(3 marks)*

- Boiling point
- Viscosity
- Flammability

> As the size of the hydrocarbon molecules increase:
> boiling point increases ✓
> viscosity increases ✓
> flammability decreases ✓.

Viscosity is a measure of the 'runniness' or thickness of a liquid. The more **viscous** it is, the less runny it is. Molecules of hydrocarbons vaporise before they burn. As bigger molecules have higher boiling points they vaporise less easily and so are less flammable.

Example

Use words from the box to complete the sentences about cracking. Each word can be used once, more than once, or not at all. *(4 marks)*

alkanes	alkenes	catalysts	fuels	gases

Cracking is the thermal decomposition of _____.

These molecules are heated until they vaporise and are either mixed with steam and heated to a very high temperature or passed over hot _____.

The products of cracking usually include **unsaturated** hydrocarbons called _____.

Small **saturated** hydrocarbons are used as _____.

Organic Chemistry

alkanes ✓
catalysts ✓
alkenes ✓
fuels ✓

You need to know why cracking of alkanes is done and in general terms how steam and catalytic cracking are carried out.

Example

Many alkanes are used as fuels. Octane, C_8H_{18}, is one of the main components of petrol. During the combustion of fuels, carbon and hydrogen are oxidised.

a) What are the products of complete combustion of octane? *(2 marks)*

Carbon dioxide ✓ and water ✓.

b) Write a balanced equation for the complete combustion of octane. State symbols are not required. *(2 marks)*

$C_8H_{18} + 12.5O_2 \rightarrow 8CO_2 + 9H_2O$

Multiples, e.g. $2C_8H_{18} + 25O_2 \rightarrow 16CO_2 + 18H_2O$, will be allowed.

Correct species ✓

Balanced ✓

c) Explain why the combustion of fuels can be considered as an oxidation reaction. *(2 marks)*

Oxidation is the gain of oxygen ✓. Both carbon and hydrogen gain oxygen during the complete combustion of hydrocarbons ✓.

Combustion means burning. Whenever a hydrocarbon is completely burnt / combusted then carbon dioxide and water are formed. Oxidation in this question is referring to the gain of oxygen.

Example

The diagram below shows how cracking can be carried out in a school laboratory.

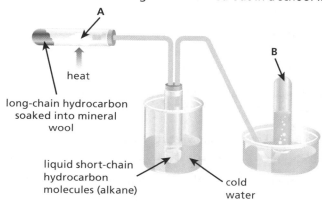

a) State the purpose of the substance labelled **A**. *(1 mark)*

> Catalyst / to speed up the reaction ✓.

b) Explain why cracking is a useful process. *(4 marks)*

> Short-chain alkanes are more useful than longer-chain alkanes ✓ for use
> as fuels and the manufacture of other chemicals ✓.
> There are more long-chain alkanes than short-chain alkanes ✓.
> Cracking is therefore useful because it converts the relatively useless long-
> chain hydrocarbons into shorter more useful ones ✓.

> Detail is important here. As well as outlining what cracking is (the breakdown of long-chain
> hydrocarbons into shorter more useful ones) the question asks why this is important and so
> you are guided towards giving more details about the uses of the hydrocarbons.

c) When decane, $C_{10}H_{22}$, is cracked the following reaction can occur.

$$C_{10}H_{22} \rightarrow C_6H_{14} + 2X$$

What is the formula of product X? *(1 mark)*

> C_2H_4 ✓

> Note that **two** molecules of X are formed. A common error in this question is to give
> C_4H_8 as an answer.

Organic Chemistry

d) From the equation in **c)**, state which substance is most likely to collect at point **B** on the diagram. *(1 mark)*

X/C_2H_4 ✓

Answer from part **c)** will be allowed if it is an alkene.

e) Suggest a use for product X. *(1 mark)*

In the manufacture of polymers / and as starting materials for the production of many other chemicals ✓.

f) When decane is cracked the following reaction can also occur.

$$C_{10}H_{22} \rightarrow C_5H_{12} + C_5H_{10}$$

Suggest why more than one possible reaction can occur when decane is cracked. *(2 marks)*

There are lots of carbon–carbon bonds in a long molecule such as decane ✓ and so the molecule can break at different positions in the chain ✓.

Think of cracking as the breaking of long chain molecules into smaller ones. You can break the molecule in many different places meaning that many different products can be formed.

g) In a separate experiment, a hydrocarbon is cracked that produces only propane and ethene as the products. Describe a test to distinguish between propane and ethene and give the result of the test. *(3 marks)*

Add bromine water ✓

Ethene will decolourise bromine water / turn it from orange to colourless ✓.

Propane will have no effect on bromine water / it remains orange ✓.

Bromine water turns colourless / is decolourised. 'Colourless' is not the same as 'clear', which is an incorrect answer frequently given to this question!

For more on the topics covered in this chapter, see pages 136–139 of the *Collins AQA GCSE Combined Science Revision Guide*.

This is another topic that lends itself to many practical activities. There could be questions that test your knowledge of practical activities, e.g. chromatography. You will have to be familiar with how this topic's practical activities can be carried out, and also how to interpret and analyse the results. Testing for gases is also covered in this topic.

Purity, Formulations and Chromatography

Chemically pure materials consist of a single element or compound and **formulations** are carefully prepared mixtures of substances in precise ratios to give the material particular properties. Chromatography is one way of separating mixtures of substances in the laboratory and this can easily be done with coloured materials such as inks.

Example

Which one of the following substances is chemically pure?
Tick **one** box. *(1 mark)*

Sodium chloride ☐

Air ☐

Crude oil ☐

Petrol for use in cars ☐

Sodium chloride ✓

A pure substance is defined as a single element or compound that is not mixed with any other substance. Petrol is a formulation. Air and crude oil are mixtures.

Chromatography

Chromatography can be used to separate mixtures and can give information to help identify substances. You will have carried out a practical involving paper chromatography and calculated R_f values.

Chemical Analysis

Example

A student used chromatography to separate the different dyes in food dyes. He sets up the apparatus as shown below.

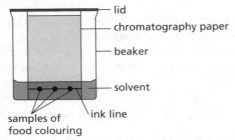

Identify **two** errors that the student made in setting up the experiment.   (2 marks)

The line should be drawn in pencil / not drawn in ink ✓.

The level of solvent should be below the ink line ✓.

The ink will dissolve in the water or if the solvent level was below the line (as it should be) then the chemicals in the ink would also rise up the chromatography paper.

Example

A chromatography experiment was carried out to investigate the different dyes present in a new brand of ink, brand 'X'. Brand 'X' ink was compared against samples of red, blue and green inks.

The final chromatogram is shown below.

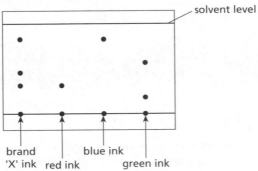

a) How many different compounds are present in brand 'X'?
    Explain your answer.   (2 marks)

Three ✓. There are three separate spots ✓.

b) Which **two** colours are known to be present in brand X? Explain how the chromatogram confirms this. *(2 marks)*

> Red and blue ✓. The R_f value / height of the spots is the same level as the spots for the red and blue inks ✓.

c) Are there are any other materials in brand 'X' ink? Explain your answer. *(2 marks)*

> There is one other substance / compound ✓. The height of the spot / R_f value does not match any of the reference inks ✓.

d) The spot for the red ink travelled 2.4 cm. The solvent travelled 7.0 cm. Calculate the R_f value for the red ink. *(2 marks)*

> $2.4 \div 7.0 = 0.34$
>
> Working ✓
>
> Answer ✓ – this mark will not be awarded if the answer is given to more than two significant figures.

e) Is the green ink pure? Explain your answer. *(2 marks)*

> No ✓ because it contains two substances ✓ (as shown by the presence of two separate spots).

f) Explain how R_f values can be used to identify the components of the green ink. *(2 marks)*

> Each compound has its own unique R_f value ✓, which can be checked / cross-referenced on a database ✓.

g) Explain how this method of chromatography separates mixtures. *(2 marks)*

> Substances in the mixture being tested are attracted to both the **stationary phase** (the chromatography paper) and the **mobile phase** (the solvent rising up the paper) ✓.
>
> The greater the attraction the substance has for the solvent, the further up the chromatography paper it will travel ✓.

Chemical Analysis

h) Brand X is a formulation. Explain the meaning of the word formulation and how a formulation is made. *(2 marks)*

> A formulation is a mixture that has been designed as a useful product ✓.
>
> It is made by mixing the components in carefully measured quantities ✓.

Pure substances will only have one spot (due to there only being one substance in them). Each substance has its own R_f value (a measure of how far up the chromatography paper it travels relative to how far the solvent has travelled).

You could be asked to calculate the R_f value in an exam by measuring the distances yourself.

Identification of Gases

You need to know the tests and observations for hydrogen, chlorine, oxygen and carbon dioxide gases.

Example

Match together the gases with their correct tests and results by drawing lines linking the boxes. *(2 marks)*

Gas	Test	Result
Hydrogen	Lit splint	Relights
Chlorine	Lime water	Turns milky / cloudy
Carbon dioxide	Glowing splint	Bleached
Oxygen	Damp litmus paper	Squeaky pop

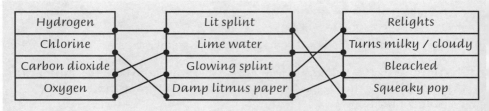

All four correct ✓ ✓
Two correct ✓

You specifically need to know the tests and results for these four gases.

For more on the topics covered in this chapter, see pages 140–141 of the *Collins AQA GCSE Combined Science Revision Guide*.

16 Chemistry of the Atmosphere

Some changes to the atmosphere are as a consequence of our actions whereas others occur as part of natural variations and cycles. Scientists try to model these changes to understand the composition of the atmosphere and to predict the environmental outcomes of these changes.

The Composition and Evolution of the Earth's Atmosphere

The composition of the atmosphere is constantly changing. As a global community, we are becoming increasingly concerned about the atmosphere and the impact that human activity is having on it. Changes to the atmosphere could have long term catastrophic consequences. Learning how the atmosphere has evolved helps us to understand some of the chemistry of the atmosphere.

Example

One theory about the composition of the Earth's early atmosphere is that intense volcanic activity released gases that formed the early atmosphere and water vapour that ultimately formed the oceans.

The composition of the atmosphere at this point may have been similar to the composition of the atmosphere of Mars today.

The pie charts below show the composition of the atmospheres of Earth and Mars today.

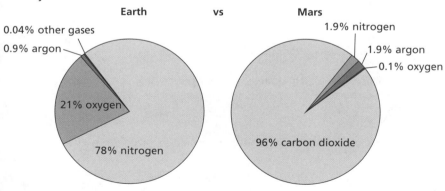

Earth vs **Mars**

Earth:
- 0.04% other gases
- 0.9% argon
- 21% oxygen
- 78% nitrogen

Mars:
- 1.9% nitrogen
- 1.9% argon
- 0.1% oxygen
- 96% carbon dioxide

Percentages do not add up to 100% due to rounding.

Chemistry of the Atmosphere

a) Which is the most abundant gas in Mars's atmosphere? *(1 mark)*

> Carbon dioxide ✓

b) Explain why the composition of the Earth's early atmosphere is only a theory. *(2 marks)*

> No data exists from that time ✓ because no people were there to record it ✓.

c) Name two gases that are present in 'other gases' in the atmosphere of Earth. *(2 marks)*

> Carbon dioxide ✓ water vapour ✓
> Any other noble gas will be allowed.

d) Suggest how water vapour in the air could lead to the formation of oceans. *(2 marks)*

> As the Earth / atmosphere cooled ✓ water vapour condensed ✓.

Pie charts are commonly used to represent data about composition of the atmosphere. In this question, you need to be able to compare the two different sets of data and draw conclusions from it before applying your knowledge of the Earth's atmosphere.

Example

Since **algae** started producing oxygen approximately 2.7 **billion** years ago, the amount of oxygen in the atmosphere gradually increased to a level that enabled animals to evolve.

a) What is the approximate percentage of oxygen in the atmosphere today? *(1 mark)*

> 21% ✓
> Answers within the range 20–22% will be allowed.

b) Name the process by which algae and other plants produce oxygen. *(1 mark)*

> Photosynthesis ✓

c) Write a balanced symbol equation for this reaction. State symbols are not required. *(2 marks)*

> $6CO_2 + 6H_2O \rightarrow C_6H_{12}O_6 + 6O_2$
> Correct formulae ✓
> Correctly balanced ✓

This question looks at how the amount of oxygen in the atmosphere has changed over time. You need to know the balanced symbol equation for **photosynthesis**.

Carbon Dioxide and Methane as Greenhouse Gases

The greenhouse effect leading to global warming is a topical issue and exam questions could test not only your knowledge of the causes and effects of global temperature change but also ask you to analyse and interpret data.

Example

The graph below shows how the amount of carbon dioxide in the atmosphere has changed since the year 1700.

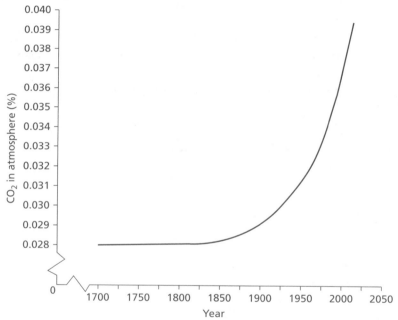

a) Suggest why the amount of carbon dioxide in the atmosphere has significantly increased from 0.028% to the levels recorded today. *(2 marks)*

Due to increased use / combustion / burning ✓ of fossil fuels ✓.

Since the Industrial Revolution the use of fossil fuels has significantly increased, particularly since the advent of motor vehicles. Nature, to a certain extent, can maintain a constant level of carbon dioxide in the atmosphere but this has not been the case in recent years – as the graph shows.

Chemistry of the Atmosphere

b) What was the percentage of carbon dioxide in the atmosphere in the year 1925? *(1 mark)*

> 0.030% ✓

c) Describe **two** ways that the environment naturally reduces the amount of carbon dioxide in the atmosphere. *(2 marks)*

> ✓✓ Two from:
>
> photosynthesis
>
> dissolving in the oceans
>
> in the formation of sedimentary rocks / fossil fuels that contain carbon.

d) Suggest why the data shown in this graph causes many climate scientists to be concerned. *(2 marks)*

> The rate at which the amount of carbon dioxide in the atmosphere is increasing is so rapid ✓ and it is believed that increasing carbon dioxide in the atmosphere is harmful ✓.

Example

Which one of the following components of air is **not** a **greenhouse gas**?
Tick **one** box. *(1 mark)*

Carbon dioxide ☐

Methane ☐

Nitrogen ☐

Water vapour ☐

> Nitrogen ✓
>
> Carbon dioxide is a commonly known greenhouse gas. You need to know that methane and water vapour are also greenhouse gases.

Example

Carbon dioxide is a gas that causes the greenhouse effect.

Describe what is meant by the term 'greenhouse effect'. Your answer should make reference to the interaction of short and long wavelength radiation with matter. *(5 marks)*

✓✓✓✓ Five from:

greenhouse gases, e.g. carbon dioxide, let short wavelength radiation, e.g. uv light, from the sun pass through

this radiation is absorbed by the surface of the Earth causing it to heat up

as the Earth cools it emits longer wavelength (infrared) radiation

greenhouse gases absorb this **infrared** radiation

this causes the bonds in these molecules to vibrate / bend / stretch more vigorously / raises their temperature

this process traps some of the heat given off by the cooling Earth in the atmosphere, leading to a higher temperature.

You need to be able to describe the greenhouse effect specifically with reference to the interaction of short and long wavelength radiation with matter.

Example

State **two** ways in which human activity increases the amount of each of the following gases in the atmosphere:

i)   Carbon dioxide

ii) Methane   *(4 marks)*

i)   Increased use of fossil fuels ✓; deforestation ✓

ii) Increased animal farming ✓; decomposition of rubbish / use of landfill sites ✓

Learn these specific examples!

Example

Based on peer-reviewed evidence, many scientists believe that human activities will cause the temperature of the Earth's atmosphere to increase at the surface and that this will result in global climate change.

a)   What is meant by the term 'peer-reviewed evidence'?   *(1 mark)*

Work / evidence of scientists that has been checked by other scientists to ensure that it is accurate and scientifically valid ✓.

b) State **three** potential effects of global climate change. *(3 marks)*

> ✓✓✓ Three from:
>
> rising sea levels, which may cause flooding and coastal erosion
>
> more frequent and severe storms
>
> changes to the amount, timing and distribution of rainfall
>
> temperature and water stress for humans and wildlife
>
> changes to the distribution of wildlife species
>
> changes in the food-producing capacity of some regions.

As well as understanding the causes of global warming and how global warming occurs, the potential effects and consequences of global warming need to be known.

Evidence for Global Warming

It is very difficult to model complex systems such as global warming, partly because there are so many variables to consider and so much data to analyse. Some people argue that the evidence for global warming is not conclusive and that the claims of scientists are not sufficient proof that it is taking place.

Example

Consider the graph below.

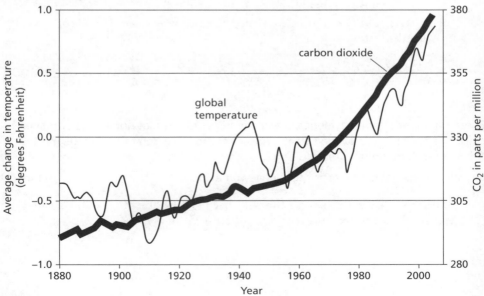

A scientist who saw this graph made the following statement:

'The graph shows that increased carbon dioxide in the atmosphere leads to increased global temperatures.'

Give one piece of evidence that supports this statement and one piece of evidence that does not support this statement. *(2 marks)*

Supporting evidence: There is a general correlation / trend between the amount of carbon dioxide in the atmosphere and the global temperature ✓.

Non-supporting evidence: The data doesn't match perfectly, i.e. there are some times when the amount of carbon dioxide in the atmosphere has increased but the global temperature has decreased and vice-versa ✓.

Questions could ask you to evaluate the validity of data, draw conclusions from it or comment on conclusions that others have made. It is therefore important that you look critically at the data and at the wording of any supporting commentary.

Example

One potential consequence of an increase in the average global temperature is rising sea levels.

Discuss the environmental implications of rising sea levels. *(3 marks)*

Rising sea levels may cause flooding ✓ and coastal erosion ✓. Rising sea levels would reduce the amount of land available for places to live / to grow crops / graze animals ✓.

You should be able to describe four potential effects of global climate change and discuss the scale, risk and environmental implications of global climate change.

Example

An advert encouraging readers to reduce their **carbon footprint** includes the following advice:

- walk short journeys instead of driving

- only put enough water in the kettle for what you need

- eat less meat.

Chemistry of the Atmosphere

a) What is meant by the term 'carbon footprint'? *(2 marks)*

> The total amount of carbon dioxide and other greenhouse gases ✓ emitted over the full life cycle of a product, service or event ✓.

b) Describe how each of the points of advice can help readers to reduce their carbon footprint. *(3 marks)*

> Walking instead of taking a car journey / using only the required amount of water in a kettle reduces the amount of fossil fuels needed ✓ and therefore reduces the amount of carbon dioxide produced ✓.
>
> Eating less meat reduces emissions of methane ✓ (methane is produced as a result of digestion and decomposition of waste).

c) Give an example of a 'carbon offsetting' measure that can be taken. *(1 mark)*

> Planting trees / investing in renewable energy ✓ are examples of carbon offsetting.

d) State **two** problems that can be encountered whilst trying to reduce the carbon footprint. *(2 marks)*

> ✓✓ Two from:
>
> disagreement over the causes and consequences of global climate change
>
> lack of public information and education
>
> lifestyle changes, e.g. greater use of cars / aeroplanes
>
> economic considerations, i.e. the financial costs of reducing the carbon footprint
>
> incomplete international co-operation.

> You need to be able to describe actions to reduce emissions of carbon dioxide and methane, and reasons why these actions may be limited.

Common Atmospheric Pollutants and their Sources

The combustion of fuels is a major source of atmospheric pollutants. You need to know these pollutants, how they are formed and the environmental and health consequences of their presence in the atmosphere.

Example

The data below shows the composition of exhaust gases from a new petrol car.

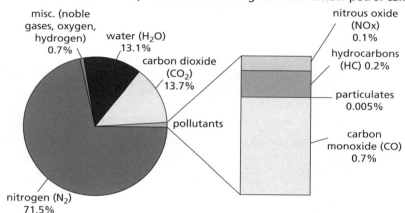

misc. (noble gases, oxygen, hydrogen) 0.7%

water (H_2O) 13.1%

carbon dioxide (CO_2) 13.7%

pollutants

nitrogen (N_2) 71.5%

nitrous oxide (NOx) 0.1%

hydrocarbons (HC) 0.2%

particulates 0.005%

carbon monoxide (CO) 0.7%

a) Explain why nitrogen is the most abundant gas in the exhaust gas mixture. *(2 marks)*

> *Nitrogen is the most abundant gas in the air* ✓ *and is not very reactive* ✓ *and so passes through the car largely unchanged.*

b) Explain why carbon dioxide and water are present in exhaust gases. *(2 marks)*

> *Small solid particles, e.g. soot / carbon, present in exhaust gases due to complete combustion* ✓ *of hydrocarbons* ✓.

A similar pie chart for a car that was 15 years old showed a higher proportion of carbon monoxide, unburnt hydrocarbons and particulates.

c) What are particulates? *(1 mark)*

> *Small solid particles present in exhaust gases* ✓.

Chemistry of the Atmosphere

d) State **one** environmental consequence of particulates. *(1 mark)*

Global dimming / health problems, e.g. respiratory problems ✓.

e) How are oxides of nitrogen produced in a car engine? *(1 mark)*

The high temperatures / sparks from spark plugs cause nitrogen and oxygen from the air to react with each other ✓.

f) Explain why, in an older car, you might expect a higher proportion of carbon monoxide and unburnt hydrocarbons. *(2 marks)*

Engines are less efficient / less oxygen is able to combust the fuel / petrol / hydrocarbons ✓, meaning that incomplete combustion will occur ✓.

g) Until 2002 most petrol in the UK contained significant quantities of sulfur.

Explain the environmental and health consequences of sulfur in petrol. *(3 marks)*

Sulfur burns to form sulfur dioxide ✓.
Sulfur dioxide can cause acid rain ✓.
Respiratory problems ✓.

You need to be able to describe how carbon monoxide, soot (i.e. carbon particles), sulfur dioxide and oxides of nitrogen are produced by burning fuels. In addition, you need to be able to describe the problems caused by increased amounts of these pollutants in the air.

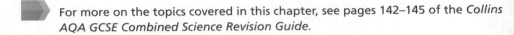 For more on the topics covered in this chapter, see pages 142–145 of the *Collins AQA GCSE Combined Science Revision Guide*.

17 — Using Resources

Many of the Earth's natural resources are used to manufacture products that are used every day and improve the quality of our lives. However, the extraction, production and disposal of many items has environmental costs. Therefore, chemists play an important role in reducing the environmental impact.

Using the Earth's Resources and Obtaining Potable Water

The Earth's natural resources are finite and, with an ever-increasing population, there is greater demand being placed on them. The potential biggest concern is for ensuring sufficient **potable** water, i.e. water that is safe to drink.

Example

For many years Brazil has been producing ethanol for use as a fuel in cars. Brazil's climate allows for sugarcane to be grown rapidly. The sugar is then fermented to produce ethanol. The ethanol is usually mixed with other fuels before being used in cars. In the UK, the majority of cars are powered using petrol or diesel. Petrol and diesel are obtained from crude oil.

a) Crude oil is a finite resource. Explain the meaning of the term 'finite'. *(1 mark)*

It [crude oil] will eventually run out at the current rate of usage / it is being used at a greater rate than it is being produced ✓.

b) Is ethanol a finite resource? Explain your answer. *(1 mark)*

No – as it can be produced by the fermentation of sugars ✓.

c) Suggest why ethanol production as a fuel for cars in the UK is not likely to be as successful as in countries such as Brazil. *(2 marks)*

The UK does not have the climate ✓ or land space available for the production of sufficient quantities of sugarcane ✓.

Using Resources

A recent estimate suggested that there were 1688 billion barrels of oil reserves in the Earth. Another estimate places the amount of oil consumed per year to be 34 billion barrels.

d) At the current rate of consumption, to the nearest year, how long will it take for the oil in the Earth to run out? *(1 mark)*

> $1688 \div 34 = 50$ years (to the nearest year) ✓

e) Explain why this figure is unlikely to be accurate. *(2 marks)*

> It is difficult to accurately predict how much oil there is / whether new reserves will be discovered ✓ and how much oil we will use in the future ✓.

> In questions such as this one you are required to apply your knowledge of the production of ethanol to the context of the question. The latter part of the question involves you being able to appreciate that the figures are estimates and why they cannot be more accurate.

Example

Most potable water in the UK is obtained from rainwater that collects in the ground as well as from water in rivers and lakes. This water is then filtered and sterilised before being pumped to your home or school.

a) What is meant by the term 'potable water'? *(1 mark)*

> Water that is safe to drink ✓.

b) In the chemical sense, is potable water pure? Explain your answer. *(2 marks)*

> No ✓ as it has other chemicals / minerals and ions dissolved in it ✓.

c) Name **two** sterilising agents used in water treatment. *(2 marks)*

> ✓✓ Two from:
> chlorine
> ozone
> UV light.

In many countries sources of potable water are scarce. Potable water can be obtained from salty water. The apparatus below can be used to **desalinate** salty water in the laboratory.

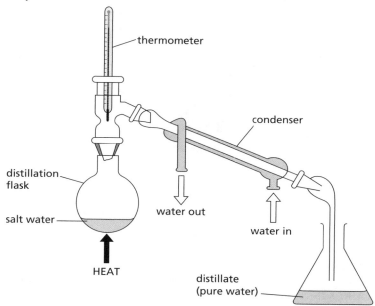

d) Name this method of removing salt from salty water. *(1 mark)*

Distillation ✓

e) What temperature will be recorded on the thermometer during this process? Explain your answer. *(2 marks)*

100°C ✓ because only pure water / steam is collected / passes over the thermometer / enters the condenser ✓.

f) Describe and explain the appearance of the inside of the distillation flask at the end of the experiment. *(2 marks)*

There will be a white residue / solid / scum / layer present ✓ due to the salt / impurities in the water ✓.

Using Resources

Another method of desalination is a process known as reverse osmosis. A diagram showing the principle behind reverse osmosis is shown below.

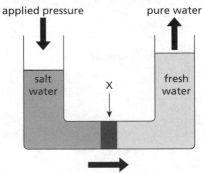

direction of water flow

g) What is represented by X? *(1 mark)*

(semi-permeable / selectively permeable) membrane ✓

h) Explain why the salt dissolved in water does not pass through. *(2 marks)*

The ions / particles present in salt ✓ are too large to pass through the gaps (in the membrane) ✓.

i) Why are methods such as that in part **d)** and reverse osmosis not normally used to produce potable water? *(1 mark)*

They require large amounts of energy that are difficult / expensive to generate ✓.

You need to be able to distinguish between potable and pure water and to be able to describe the differences in how ground water and salty water are treated to make it potable.

Example

The flow chart below shows how sewage can be treated.

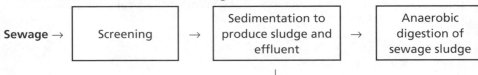

a) At which stage of the sewage treatment process are large items such as twigs removed? *(1 mark)*

Screening ✓

b) Describe what happens during the sedimentation stage. *(1 mark)*

The screened sewage is allowed to settle and the solid particles fall to the bottom ✓ of the tank to form a sediment.

c) What is the difference between aerobic and anaerobic conditions? *(1 mark)*

Aerobic is in the presence of air / oxygen and anaerobic is in the absence of air / oxygen ✓.

d) Describe the role that bacteria play in the treatment of sewage. *(2 marks)*

Bacteria feed on remaining organic material and harmful microorganisms ✓ breaking them down into less harmful materials ✓.

Make sure you know the four different stages of sewage treatment and can give a brief explanation of what happens at each stage.

HT Example

Read the extract below from an article about copper mining and then answer the questions that follow.

Traditional mining or quarrying for metals involves digging, moving and disposing of large amounts of rock. This is both bad for the environment and, with the remaining mines containing ores with less and less copper in them, the costs of mining for copper mean that this is less financially viable for the mining companies. In recent years many copper mines have stopped production. Given that the demand for copper remains high, and is likely to increase further, new methods of obtaining copper from low-grade ores, such as **bioleaching**, are being developed.

a) Why might traditional mining be described as 'bad for the environment'? *(1 mark)*

✓ One from:

pollution caused by dust / noise

pollution caused by traffic / lorries to / from the quarry

physical damage to the land, e.g. the mine / quarry itself or in building the road networks.

b) What is an ore? *(1 mark)*

> A rock that contains metal in high enough quantities to make it economically worthwhile to extract ✓.

c) What is copper used for and why might the demand for copper increase further? *(3 marks)*

> ✓✓ Any two from:
>
> water pipes / building, e.g. roofs / electrical wiring / making alloys, e.g. brass / making saucepans / cooking equipment ✓✓.
>
> ✓ Any one from:
>
> Increased demand for / development of electrical items will lead to an increased demand for copper to be used in circuits / wiring ✓.

d) Describe how the process of bioleaching enables copper to be obtained. *(2 marks)*

> Bacteria are used to extract metal from low-grade ores forming a solution known as a leachate ✓.
>
> The copper is then obtained from the leachate, e.g. by displacement or electrolysis ✓.

With increasing focus on environmental issues and sustainability the exam will cover some of the chemistry and general issues around these topics. As well as having a knowledge of processes such as bioleaching, you need to be able to demonstrate an awareness of the environmental implications of such innovations and to be able to evaluate information presented to you.

Example

HT Some plants, such as cabbages, can absorb metal ions through their roots and store the accumulated metal. The plants can then be burnt and the metal extracted from the ash formed. This process is known as **phytomining**.

State **two** advantages and **one** disadvantage of phytomining over conventional mining. *(3 marks)*

> ✓✓ Advantages include:
>
> ability to extract metals from low-grade ores
>
> fewer quarries / mines needed, therefore preserving the environment
>
> on a large scale, phytomining is more cost-effective than mining
>
> less waste material produced.

✓ Disadvantages include:

pollution produced when the crops are burnt

lots of land is sacrificed for phytomining that could be used for other purposes

the local ecology can be affected

phytomining is unpredictable and subject to variations in weather.

Bioleaching and phytomining are two alternatives to conventional mining for metals that are specifically mentioned in the specification. So, make sure you are aware of how they work and the pros and cons of each – particularly with regards to the impact on the environment.

Life Cycle Assessment and Recycling

Life cycle assessments (LCAs) consider the environmental impact of products throughout their life, i.e. from obtaining the raw materials all the way through to their disposal.

Example

The data below shows a LCA for both paper shopping bags and polythene (plastic) bags. The raw material for paper bags is trees and for polythene bags is crude oil.

The values given refer to 1000 bags over their whole life and assumes that each item is used only once.

	Paper	Polythene (plastic)
Energy use (MJ)	2590	713
Fossil fuel use (kg)	28	13
Solid waste (kg)	34	6
Greenhouse gas emissions (kg CO_2)	72	36
Freshwater use (litres)	3387	198

a) Suggest why more solid waste is produced in the manufacture of paper bags than in the manufacture of polythene bags. *(2 marks)*

> Not all the tree is used to make paper bags ✓. All components of crude oil are used for some purpose and so there is relatively little waste produced ✓.

b) Why might the values attributed to the greenhouse gas emissions not be accurate? *(1 mark)*

> The values are very difficult to measure / directly attribute ✓.

c) Based on the above LCA, is using paper or polythene bags better for the environment? Explain your answer. *(2 marks)*

> *Polythene bags* ✓ *because in every category the environmental impact is less* ✓.

In 2015 the British Government introduced a charge of 5p per polythene bag obtained from supermarkets.

d) Explain how the introduction of this charge will impact on the LCA values in the table. *(2 marks)*

> *The charge encourages people to reuse / not dispose of polythene bags* ✓ *and so the number of uses per bag increases, meaning the LCA values will decrease* ✓ *(because the data assumed that each bag was used only once).*

When considering LCAs you should understand why value judgements are sometimes used. For example, allocating numerical values is not straightforward or easy to calculate. You should also be mindful that selected data from a LCA can be used to make claims, e.g. by advertisers, that might not be wholly objective.

Example

Every day over 35 million plastic bottles are sold in Britain. Some estimates show that 10 million tons of plastic end up in the world's oceans every year.

a) Suggest why the production and disposal of plastic bottles damages the environment. *(2 marks)*

> *Plastic bottles are produced from crude oil and it takes a lot of energy* ✓ *to produce plastic bottles. The disposal of plastic bottles can harm wildlife* ✓.

b) Discuss how the use of plastic can be reduced. *(3 marks)*

> ✓✓✓ *Three from:*
> *by finding alternatives to reduce the demand for plastic bottles*
> *by recycling plastic bottles*
> *by reusing plastic bottles*
> *by finding alternative uses for them after use.*

The three key words when considering how to use less resources are Reduce, Reuse and Recycle. You should be able to apply this idea to plastics, glass and metals.

For more on the topics covered in this chapter, see pages 146–149 of the *Collins AQA GCSE Combined Science Revision Guide*.

Changes in Energy

This topic first covers some underlying principles of **energy** changes before going on to cover the use of a number of energy transfer equations and practical applications of energy changes.

Example

Which statement below is **incorrect**? Tick **one** box.   *(1 mark)*

Energy can be transferred.   ☐

Energy can be created.   ☐

Energy can be stored.   ☐

There will be some multiple choice questions in the exam. This is a knowledge recall question that tests knowledge of a vitally important concept in the topic of Energy (and one that is going to be useful in future questions in this section!).

Energy can be created. ✓

As well as not being able to be created, energy cannot be destroyed either. This means all the energy in the **Universe** has existed since the start of time and will exist until the end of time; it is just transferred into different forms.

Example

Use answers from the box to complete the sentences below. Each word can be used once, more than once or not at all.   *(2 marks)*

electrical	system	gravitational	heat

A _____ is an object or a group of objects. There are changes

in the way energy is stored when a _____ changes. When an

electric kettle boils, _____ energy is being converted into

_____ and sound energy.

Energy

A **system** is an object or group of objects. There are changes in the way energy is stored when a **system** changes. When an electric kettle boils, **electrical** energy is being converted into **heat** and sound energy. ✓ ✓

One incorrect answer −1 mark

There may be 'fill in the blank' questions like this in the exam and this one seems straightforward with four words and four gaps. However, it is important to consider that the instructions say that each word can be used once, more than once or not at all. In this example one of the words isn't used at all whilst another is used twice.

When considering energy, a system is just a group of objects. Energy is stored in the objects that make up a system, so when the system changes the way the energy stored within the system also changes.

Example

A group of scientists carried out an investigation into slingshots. In Experiment 1 a ball of **mass** 15 g was fired from a slingshot. It reached a maximum **speed** of 20 m/s.

a) Write down the equation that relates **kinetic energy**, mass and speed. *(1 mark)*

There are certain equations that you need to recall and some that will be given on the Physics Equations Sheet in the exam. This equation is an example of one which you need to recall. It is very important that you learn these equations thoroughly as there may be a mark for correctly writing down the equation and then marks for applying the equation to the question.

kinetic energy = 0.5 × mass × (speed)² ✓

b) Calculate the maximum kinetic energy of the ball. *(2 marks)*

Questions will often not just be a straightforward substitution into a formula. In this case the kinetic energy equation uses kg as a unit of mass, but the question gives the mass in grams. You therefore need to convert the mass into kg before substituting the values into the equation.

$$15\,g = 0.015\,kg$$
$$\text{kinetic energy} = 0.5 \times 0.015 \times (20)^2$$
$$= 0.5 \times 0.015 \times 400 \checkmark$$
$$= 3\,J \checkmark$$

c) **i)** In Experiment 2, the elastic of the same slingshot was pulled back 0.35 m. The **elastic potential energy** of the slingshot elastic was 2.45 J and the **limit of proportionality** was not exceeded.

Determine the spring constant of the slingshot elastic.

Use the correct equation from the Physics Equations Sheet. *(3 marks)*

> The formula for elastic potential energy will be given to you in the Physics Equations Sheet in the exam paper (see pages 302–303 of this book). You could be asked to change the subject of an equation in the exam and this question requires you to do so in order to find the spring constant. In all calculations it is very important to show your working. However, it is particularly important when changing the subject of an equation as there will usually be a mark for the correct rearrangement, the correct substitution and the correct answer. Examiners will award the full mark allocation if you just write the correct answer, but this is a risky strategy because a wrong answer will score no marks whilst an incorrect answer with some correct working may score some marks.

elastic potential energy = 0.5 × spring constant × (extension)^2

spring constant = elastic potential energy ÷ (extension^2 × 0.5) ✓

= 2.45 ÷ (0.35^2 × 0.5)

= 2.45 ÷ (0.1225 × 0.5) ✓

= 40 N/m ✓

ii) Was the **extension** of the slingshot greater in Experiment 1 or Experiment 2?

Explain your answer. *(2 marks)*

> The exam papers will include lots of questions that ask you to apply your physics knowledge to experimental examples. In this case you have to deduce the extension of the slingshot in Experiment 1. The key to this question is the fundamental idea that energy cannot be created. The ball in Experiment 1 had a maximum kinetic energy of 3 J, so the elastic potential energy of the slingshot elastic must have been 3 J or greater in order to transfer 3 J of energy to the ball. This is greater than the elastic potential energy in Experiment 2. As the same slingshot was used in both investigations, the spring constants must have been the same. Therefore the only way to give a greater elastic potential energy is to have a greater extension.

The extension must have been greater in Experiment 1 ✓ as this must have had a greater elastic potential energy than in Experiment 2 in order to transfer 3 J of energy to the ball. ✓

Energy

Example

A plane is flying at a constant altitude (height) of 800 m. It begins its descent to a landing area in a field, lands and then moves along the field for a short distance.

a) Explain how the **gravitational potential** energy of the plane changes in the above situation. *(3 marks)*

> The gravitational potential energy of the plane remains constant whilst its altitude remains constant. ✓ As it descends to the field, its gravitational potential energy decreases. ✓ When the plane touches down on the field, its gravitational potential energy is zero. ✓
>
> Whilst this question seems relatively straightforward, it requires you to understand that once the plane lands the height of the plane is zero, the gravitational potential energy is also zero and will remain zero whilst the plane is on the ground.

b) Calculate the gravitational potential energy of the plane when it is flying at 800 m. The mass of the plane is 650 kg and the gravitational field strength is 9.8 N/kg.

Write down any equations you use and give your answer in kilojoules to three significant figures. *(3 marks)*

> The equation for gravitational potential energy is another equation you need to recall. You will always be given the gravitational field strength in a question; on Earth it is 9.8 N/kg. In this example, you are asked to give your answer in kilojoules as it is such a large value. A thousand joules is a kilojoule. You should make sure you are confident in converting between different magnitudes of SI units.

> gravitational
> potential energy (GPE) = mass × gravitational field strength × height ✓
> GPE = 650 × 9.8 × 800 ✓
> = 5 096 000 J
> = 5100 kJ (3 s.f.) ✓

Example

The table below shows the **specific heat capacity** of different substances.

Substance	Specific heat capacity (J/kg/°C)
Copper	390
Steel	460
Water	4180
Ethanol	2440

a) Determine the change in thermal energy when 0.32 kg of ethanol is heated from 34°C to 53°C.

Use the correct equation from the Physics Equations Sheet. *(2 marks)*

First determine the temperature change before substituting values into the formula.

$$\text{change in temperature} = 53 - 34 = 19°C ✓$$
$$\text{change in thermal energy} = 0.32 \times 2440 \times 19$$
$$= 14\,835.2 \text{ J} ✓$$

b) If the same increase in thermal energy occurred in the same mass of water at 34°C, what temperature would it reach? Explain the differences in the values. *(4 marks)*

To answer this question, you need to rearrange the equation so that temperature change is the subject and then substitute the values from part **a)** and the specific heat capacity of water from the table. Then explain why the values are different.

$$\text{change in thermal energy} = mass \times specific\ heat\ capacity \times temperature\ change$$

$$\text{temperature change} = change\ in\ thermal\ energy \div (mass \times specific\ heat\ capacity) ✓$$

$$= 14\,835.2 \div (0.32 \times 4180) ✓$$

$$= 14\,835.2 \div 1337.6$$

$$\text{temperature change} = +11°C$$

$$\text{new water temperature} = 45°C ✓$$

The water temperature didn't reach as high a value as the ethanol temperature, as water has a higher specific heat capacity than ethanol so it requires a greater increase in energy to cause the same rise in temperature. ✓

Power and Conservation and Dissipation of Energy

Example

When energy transfers take place in a closed **system**, what happens to the total energy of the system? Tick **one** box. *(1 mark)*

There is a net increase in total energy. ☐

There is a net decrease in total energy. ☐

There is no net change in total energy. ☐

There is no net change in total energy. ✓

Net change means no overall change.

Example

A model crane was used in an investigation into **power**. The electrical motor on the crane was used to raise a **mass** of 0.5 kg to a height of 1.3 m in 7 seconds.

a) i) The **force** exerted by the crane is equal to the **weight** of the mass.

Calculate the force exerted by the crane.
Assume the gravitational field strength is 9.8 N/kg.
Write down any equations you use. *(2 marks)*

weight = mass × gravitational field strength
weight = 0.5 × 9.8 ✓
weight = 4.9 N ✓

ii) Write down the equation that links **work** done, force and **distance** moved along the line of action of the force. *(1 mark)*

work done = force × distance ✓

The units for this equation are work done in joules (J), force in newtons (N) and distance in metres (m).

iii) Calculate the work done by the model crane. *(2 marks)*

work done = 4.9 × 1.3 ✓
work done = 6.37 J ✓

Use the weight value from the answer for **i)** as the force value.

iv) Write down the equation that links power, work done and time. *(1 mark)*

power = $\frac{\text{work done}}{\text{time}}$ ✓

The units for this equation are power in watts (W), work done in joules (J) and time in seconds (s).

v) Calculate the power of the model crane. *(2 marks)*

power = $\frac{6.37}{7}$ ✓
power = 0.91 W ✓

b) A second model crane raised the mass in 10 seconds. Compare the power of the first crane with the second crane.
Explain your answer. *(2 marks)*

As the mass is staying the same but the time taken is increasing ✓ the second crane must have a lower power than the first crane ✓.

You could answer this question by calculating the power of the second model crane. There is however no need to do this as the crane is lifting the same mass in a longer time.

c) **i)** What useful energy transfer is occurring in this experiment? *(2 marks)*

Electrical energy to kinetic energy ✓ to gravitational potential energy ✓.

As the question specifies a transfer you need to ensure you state at least **two** types of energy; the initial energy and the form it is **transferred** to. Refer back to the start of the question for the initial energy.

ii) Name **one** possible form of non-useful energy that may be released in this investigation. *(1 mark)*

> Heat or sound ✓

When asked to identify wasted energy first ensure you know what the useful energy transfer is! In a process involving moving parts there will be friction, which will release heat energy. Moving objects will also release sound energy.

d) **HT** Explain how a lubricant could be used to increase the **efficiency** of the cranes. *(2 marks)*

> Applying lubricant to moving parts would decrease the friction between the moving parts ✓. This would decrease the energy lost as heat so make the transfer more efficient ✓.

You are expected to be able to describe ways to increase the efficiency of an energy transfer. In a transfer involving moving parts lubricants would be a common way to do this.

Example

During 2 hours of operation a computer has a total energy input of 3240 kJ. It has a useful energy output of 2750 kJ.

a) Write down the equation that links efficiency, useful output energy transfer and total input energy transfer. *(1 mark)*

> $$\text{efficiency} = \frac{\text{useful output energy transfer}}{\text{total input energy transfer}} ✓$$

b) Calculate the efficiency of the computer.
Give your answer as a percentage. *(2 marks)*

> $$\text{efficiency} = \frac{2750}{3240} ✓$$
> $$\text{efficiency} = 0.85$$
> $$\text{efficiency} = 0.85 \times 100 = 85\% ✓$$

To convert the decimal to a percentage multiply it by 100.

c) An older version of the computer had the same total input energy transfer but an efficiency of 78%.
What was the useful output energy? *(2 marks)*

useful output energy transfer = efficiency × total input energy transfer
useful output energy transfer = 0.78 × 3240 ✓
useful output energy transfer = 2527 kJ ✓

As the efficiency of the older computer is lower than that of the newer computer and its total input energy is the same, then the useful output energy must be lower than 2750 kJ.

Example

An investigation was carried out into the effectiveness of different insulation materials. The three different insulation materials were wrapped around three different containers of water. 100°C water was added to each container and then the temperature of the water was recorded at regular time intervals. The table below shows the results of the investigation.

Material	Temperature (°C)			
	Start	1 minute	2 minutes	3 minutes
A	100	95	87	80
B	100	99	96	95
C	100	99	97	93

a) Calculate the average change in temperature for each material.
Give your answer in °C per minute to two significant figures. *(3 marks)*

$A = \frac{(80 - 100)}{3}$
$A = \frac{-20}{3}$
$A = -6.7°C / minute$ ✓

$B = \frac{(95 - 100)}{3}$
$B = \frac{-5}{3}$
$B = -1.7°C / minute$ ✓

$C = \frac{(93 - 100)}{3}$
$C = \frac{-7}{3}$
$C = -2.3°C / minute$ ✓

To calculate the average change in temperature, take the final temperature away from the start temperature and divide by the time taken.

Energy

b) Which of the three materials had the highest thermal **conductivity**?
Explain your answer using the results from **a)**. *(2 marks)*

> Material A ✓, as this showed the largest average decrease in temperature per minute ✓.

c) Why was it important that the water in all three containers had the same initial temperature? *(1 mark)*

> To ensure the temperature changes can be compared with the same start point ✓.

> In questions like this you should avoid writing 'to make it a fair test'.

d) The experiment was repeated with exactly the same conditions but in a colder room. How would the results differ? *(2 marks)*

> The temperatures would decrease to lower levels ✓ as there is a greater difference between the room temperature and the initial temperature of the water. This would lead to a greater average temperature change per minute ✓.

Example

a) Explain why a homeowner would want to reduce the thermal conductivity of the walls of their home. *(2 marks)*

> Reducing thermal conductivity reduces heat loss from the house ✓, which will mean less energy is required to keep the house warm in the winter, which will save money on energy bills ✓.

b) A homeowner's heating system has a useful power output per day of 1.8 kW. The total power input per day is 2.1 kW.

i) Write down the equation that links efficiency, useful power output and total power input. *(1 mark)*

> $$\text{efficiency} = \frac{\text{useful power output}}{\text{total power input}}$$ ✓

ii) Calculate the efficiency of the heating system.
Give your answer as a percentage. *(2 marks)*

> $\text{efficiency} = \frac{1.8}{2.1}$ ✓
> $\text{efficiency} = 0.86 \times 100 = 86\%$ ✓

iii) A new boiler increases the efficiency of the heating system to 91% whilst maintaining the same total power input.
What is the new useful power output per day? *(2 marks)*

> useful output energy transfer = efficiency × total input energy transfer
> useful output energy transfer = 0.91 × 2.1 ✓
> useful output energy transfer = 1.91 kW per day ✓

Convert the percentage efficiency into a decimal before substituting into the rearranged equation.

National and Global Energy Resources

This topic focuses on energy resources, comparing **renewable** and non-renewable energy resources, including their environmental impact, changing trends in their use and their suitability in different situations.

Example

a) Complete the table using the words in the box below. *(2 marks)*

| wind | wave | oil | tidal | natural gas |

Renewable	Non-renewable

Renewable	Non-renewable
wind	oil
wave	natural gas
tidal	
✓	✓

One mark for each correct column.

Renewable energy resources are those that can be replenished as they are used. All fossil fuels are examples of non-renewable energy resources.

Energy

b) A hospital is considering installing solar panels to provide electricity from a renewable source.

i) Give **one** advantage of solar panels over generating electricity using fossil fuels. *(1 mark)*

> Solar does not produce greenhouse gases such as carbon dioxide, whilst burning fossil fuels does ✓.

Solar is mentioned in the question as a renewable energy resource so you cannot use this as an advantage. You should also avoid subjective, simplistic statements such as 'cheap'. Instead, consider what is the big advantage of solar energy over fossil fuels.

ii) Explain why it would be dangerous for the hospital to get all of its electricity from solar panels. *(2 marks)*

> Solar panels do not provide a constant source of electricity ✓, which is required for the safe running of the hospital ✓.

In the exam you may be asked to comment on the appropriateness of different energy resources for different uses. In this case the issue is the reliability of solar panels.

Example

The table below shows how the UK's total energy consumption is divided between different sectors.

Sector	Energy consumption (%)
transport	40
industry	14
services	17
domestic	

a) Calculate the energy consumption of the domestic sector. *(1 mark)*

> domestic = 100 − 40 − 17 − 14 = 29% ✓

As this data refers to total energy consumption, the percentages must add up to 100.

b) **i)** The UK consumed 90.73 terajoules (TJ) of energy during the year. What is this in joules? Give your answer in standard form to two significant figures. *(1 mark)*

> The prefix 'tera' denotes multiplying by 10^{12}.

> 90.73×10^{12} J = 9.1×10^{13} J ✓

ii) Determine the energy used by services during the year. Give your answer in terajoules (TJ) to two significant figures. *(2 marks)*

> To answer this question, multiply the total UK energy by the percentage used by the services sector.

> Energy used by services sector = 90.73 × 17%
>
> = 90.73 x 0.17 ✓
>
> = 15 TJ ✓

c) The domestic sector mainly uses natural gas and electricity generated from both renewable and non-renewable energy resources. Explain how the energy resources used in transport may differ from this. *(1 mark)*

> Transport mainly uses fuels derived from oil and a small amount of electricity ✓.

> You need to be able to compare the way different energy resources are used in electricity generation, transport and heating.

d) How do you predict that the energy resources used in transport and the domestic sector will change over the next 50 years? *(2 marks)*

> In the domestic sector there will probably be an increase in the amount of electricity from renewable energy resources ✓. In transport there will probably be a decrease in the use of fuels derived from oil and an increase in the use of electricity ✓.

> You could be asked to consider trends and patterns in energy use; think particularly about how the balance between non-renewable and renewable energy resources has changed and will have to change in the future.

Example

Large batteries are being built to efficiently store electricity generated by wind farms. Explain how this development is helping to overcome some of the problems of electricity generation by wind farms.   *(4 marks)*

Wind farms only generate electricity when the wind is blowing ✓. This makes them an unreliable energy resource as they cannot supply electricity consistently ✓. Large batteries could be used to store excess electricity generated and then release it into the grid when the wind is not blowing ✓. This would mean homes and businesses could still receive a consistent supply of electricity even when the wind is not blowing ✓.

This is an example of a 4-mark extended response question. Like the 6-mark extended response questions, a 4-mark extended response question is marked in levels. The table below shows the different level descriptors. In order for your answer to be placed in a level you have to satisfy the criteria laid out for that particular level. The quality of your answer then determines what mark you are awarded within that level. The key aspect of all these questions is developing a sustained line of reasoning that is coherent, relevant, supported by evidence (substantiated) and logically structured. This means you should focus on writing well structured answers with a logical order that relate directly to the question and do not contain any non-relevant material.

Level	Marks
Level 2: Full description of how the battery overcomes the most significant disadvantage of wind power and the importance of this in producing a consistent supply to the electricity grid.	3–4
Level 1: Basic description of the advantages of a battery with little explanation of how this would benefit consistent supply to the grid.	1–2
No relevant content.	0

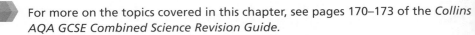

For more on the topics covered in this chapter, see pages 170–173 of the *Collins AQA GCSE Combined Science Revision Guide*.

Standard Circuit Diagram Symbols

This is a short topic that covers the different circuit symbols used to construct circuit diagrams.

Example

Draw the symbols of the following components.   *(3 marks)*

a) Lamp

b) Variable resistor

c) Diode

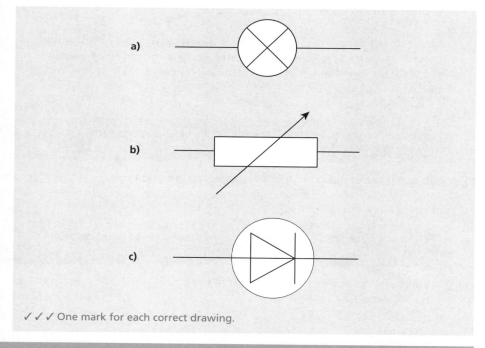

✓ ✓ ✓ One mark for each correct drawing.

You need to be able to identify, draw and interpret 14 different circuit symbols.

Electrical Charge and Current

This is a short, fairly straightforward topic but it does contain some very important equations.

Example

a) Use the answers in the box below to complete the sentence. Each word can be used once, more than once or not at all. *(2 marks)*

potential	charge	current	closed	resistance

For electrical charge to flow through a _____ circuit, the circuit

must include a source of _____ difference. Electric current is a

flow of electrical charge. The size of the electric _____ is the

rate of flow of electrical _____.

> For electrical charge to flow through a **<u>closed</u>** circuit, the circuit must include a source of **<u>potential</u>** difference. Electric current is a flow of electrical charge. The size of the electric **<u>current</u>** is the rate of flow of electrical **<u>charge</u>**. ✓ ✓
>
> −1 mark for an incorrect answer.

> In this question five options are given for four answers – this means you'll not use at least one of the words given.

b) Write down the equation that links **charge**, current and time. *(1 mark)*

> charge flow = current × time ✓
>
> The units for this equation are charge flow in coulombs, current in amperes and time in seconds.

c) i) Calculate the charge flow at a point in the circuit where the current is 6 amps for 13 seconds. *(2 marks)*

> charge flow = 6 × 13 ✓
> charge flow = 78 coulombs (C) ✓

ii) For the charge flow to change to 150 C, how long would the current have to be flowing? *(2 marks)*

$time = \dfrac{charge\ flow}{current}$

$time = \dfrac{150}{6}$ ✓

$time = 25\ s$ ✓

The charge flow is greater than the answer to **c) i)** so the time must be greater than 13 seconds.

iii) The circuit is a single closed loop. What is the current at a second point on the circuit? *(1 mark)*

6 A ✓

This question requires you to know that the current is the same in every point of a single closed loop (a series circuit).

Example

During a school physics experiment, a circuit is set up using a battery.

The circuit is a single closed loop and an ammeter in the circuit shows a reading of 3 A. The total **resistance** of the circuit is 3 ohms.

a) Write down the equation that links **potential difference**, current and resistance. *(1 mark)*

$potential\ difference = current \times resistance$ ✓

The units of this equation are potential difference in volts, current in amperes and resistance in ohms. In the exam, questions will use the term potential difference. You will gain credit for the correct use of either potential difference or voltage but it's good practice to consistently use the term potential difference.

b) Calculate the potential difference of the circuit. *(2 marks)*

$potential\ difference = 3 \times 3$ ✓

$potential\ difference = 9\ V$ ✓

c) A second battery, with the same potential difference as the original, is added to the circuit. The resistance in the circuit remains the same.
    What is the new current flowing through the circuit? *(2 marks)*

$current = \dfrac{potential\ difference}{resistance}$

$current = \dfrac{18}{3}$ ✓

$current = 6\,A$ ✓

As the potential difference has increased to 18 V whilst the resistance has remained constant, the current in the circuit will increase.

Resistors

In this topic you need to learn the shapes of the current / potential difference graphs of three resistors and be able to explain these shapes. You also need to explain how LDRs and thermistors can be used in practical applications.

Example

a) Identify the component that would produce the current / potential difference graph below and explain its shape. *(3 marks)*

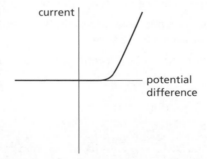

Diode ✓

The current through a diode flows in one direction only ✓ as a diode has a very high resistance in the reverse direction ✓.

This is one of three potential difference current graphs you need to learn. They each have a characteristic shape due to the nature of the component. In each case you need to explain why the component produces the graph that it does.

b) Explain why it would be wrong to describe this component as an **ohmic conductor**. *(2 marks)*

> The current through an ohmic conductor is directly **proportional** to the potential difference across the resistor ✓. This graph does not show a directly proportional relationship ✓.

> You need to learn the definition of an ohmic conductor.
> On a line graph a directly proportional relationship would be a straight line passing through the origin.

c) Explain how an LDR could be used to ensure outside lights only turn on during the hours of darkness. *(3 marks)*

> The resistance of an LDR increases as light intensity decreases ✓. The LDR should therefore be wired in a circuit that turns a light on if the current flowing through the circuit is low, as this will occur when the resistance of the LDR is high, which is in the dark ✓, and will not occur in the light, when the LDR resistance is low ✓.

Example

The table below shows the results into an investigation into the effect of temperature on the resistance in a circuit that contains a thermistor.

Temperature (°C)	Current (A)	Voltage (V)	Resistance (Ω)
20	3	8	2.7
30	6	8	Z
40	9	8	0.9

a) Calculate the missing resistance value, Z.
Show your working, including the equation used. *(3 marks)*

> $resistance = \frac{potential\ difference}{current}$ ✓
>
> $resistance = \frac{8}{6}$ ✓
>
> $resistance = 1.3\,Ω$ ✓

> This question requires you to rearrange the following equation:
> potential difference = current × resistance

b) Explain the pattern shown by the resistance values. *(1 mark)*

> As the temperature increases the resistance decreases ✓.

> When describing a trend, state the change in the independent variable (in this case temperature) and the corresponding change in the dependent variable (in this case resistance).

c) Describe how a thermistor like this one could be used in the thermostat of a heating system. *(3 marks)*

> As the temperature of the thermistor increases the resistance decreases ✓. The thermistor should be wired into a circuit that turns the heating system on when the current through the circuit is low ✓, as this will occur when the resistance of the thermistor is high in lower temperatures ✓.

Series and Parallel Circuits

This topic covers the differing properties of components wired in **parallel** and wired in **series**.

Example

The below circuit was set up during the testing of a new fuse design.

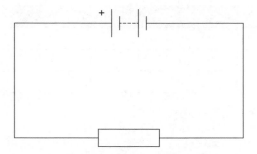

a) Identify this type of circuit. Explain your answer. *(2 marks)*

> A series circuit ✓, as the components are wired in a single closed loop ✓.

> Electrical components can be joined in series or in parallel. Some circuits will be either series or parallel but some circuits will have both series and parallel parts.

b) Explain what components would need to be added to the circuit in order to find the resistance of the fuse. *(2 marks)*

Wire an ammeter in series with the fuse to find the current drawn by it ✓ and a voltmeter in parallel to find the potential difference across the fuse ✓.

In order to calculate resistance of a component you need to find the current drawn by the component and the potential difference across it.

Example

The circuit below was built during a physics practical in a school lab.

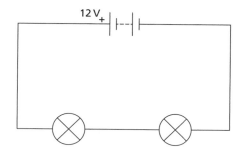

a) The resistance of each of the lamps is 3 Ω. What would be the total resistance of the circuit?
Write down any equations you use. *(2 marks)*

$R_{total} = R_1 + R_2$ ✓
$R_{total} = 3 + 3$
$R_{total} = 6\ \Omega$ ✓

In a series circuit the total resistance of two components is the sum of the resistance of each component.

Electricity

b) The circuit is modified as shown below. Each of the meters has a negligible resistance.

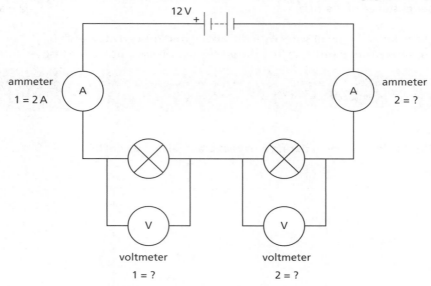

Calculate the readings that would be shown on the following meters. In each case explain how you arrived at your answer.

i) Ammeter 2 *(2 marks)*

> 2A ✓, as it would be the same reading as ammeter 1 ✓.

ii) Voltmeter 1 and voltmeter 2 *(2 marks)*

> Both would have a reading of 6V ✓, as the potential difference of the supply would be shared between each of the lamps ✓.

> In a series circuit the current is the same through each component and the total potential difference of the power supply is shared between the components. As the meters all have a negligible resistance you do not need to consider them in your answer.

Example

The circuit below was constructed during an investigation into the resistance of different components.

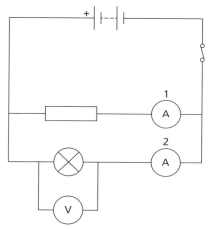

a) The voltmeter gives a reading of 9 V. What would the potential difference be across the resistor?

Explain how you arrived at your answer. *(2 marks)*

> *The potential difference across the resistor would be 9V ✓, as it would be the same as the potential difference across the lamp, as this is a parallel circuit ✓.*

In a parallel circuit the potential difference across each component is the same.

b) Ammeter 1 gives a reading of 3 A and ammeter 2 gives a reading of 6 A. What would be the total current through the whole circuit?

Explain how you arrived at your answer. *(2 marks)*

> *The total current would be 9A ✓, as the total current in a parallel circuit is the sum of the currents through the branches ✓.*

The total current in a parallel circuit is the sum of the currents through the separate branches.

c) Predict the effect of adding additional resistors in parallel to this circuit.
How would this effect be different if this was a series circuit?
In both cases explain your answer. *(4 marks)*

> Adding additional resistors in parallel would decrease the total resistance
> of this circuit ✓, as the total resistance of a parallel circuit is less than the
> resistance of the smallest individual resistor ✓. In a series circuit adding
> additional resistors increases the total resistance ✓, as the total resistance
> is the sum of the resistance of each individual resistor ✓.

Domestic Uses and Safety

For all areas of this topic it's important to explain how certain features improve safety
and reduce risk.

Example

a) The graph below shows the potential difference in a circuit.

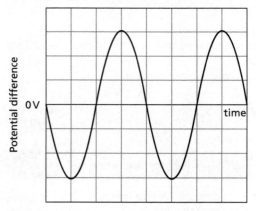

Is this an example of **a.c.** or **d.c.**?
Explain how you arrived at your answer. *(2 marks)*

> a.c. ✓, as the potential difference is alternating ✓.

b) Use the answers in the box below to complete the sentences. Each possible answer can be used once, more than once or not at all. *(2 marks)*

a.c.	Hz	230	50	d.c.	60

Mains electricity is an _____ supply. In the United Kingdom

the domestic electricity supply has a **frequency** of _____ _____

and is about _____ V.

Mains electricity is an **a.c.** supply. In the United Kingdom the domestic electricity supply has a frequency of **50 Hz** and is about **230**V. ✓ ✓
One incorrect answer –1 mark.

c) i) Complete the table of the insulation colours of the wires in the three-core cables that connect most electrical appliances to the mains. *(3 marks)*

Insulation colour	Wire
brown	
blue	
green and yellow stripes	

Insulation colour	Wire	
brown	live	✓
blue	neutral	✓
green and yellow stripes	earth	✓

One incorrect answer –1 mark.

ii) Explain why it's important that the insulation of wires is colour coded. *(2 marks)*

So they can be easily identified ✓, allowing a plug to be wired correctly. It also allows easy identification of the potentially dangerous live wire ✓.

iii) Describe the differences between the potentials of the live and the neutral wires. Explain your answer. *(2 marks)*

The potential difference between the live wire and the neutral wire is about 230V ✓ as the neutral wire is at earth potential 0V ✓.

iv) In what situation would the earth wire carry a current? *(1 mark)*

> *If there was a fault with the appliance* ✓.

Energy Transfers

This is a wide ranging topic covering power, energy transfers in everyday appliances and the **National Grid**.

Example

A games console draws a current of 5A and a resistance of 10 ohms.

a) Write down the equation that relates power, current and resistance. *(1 mark)*

> *power = (current)² × resistance* ✓
>
> The units of this equation are power in watts, current in amperes and resistance in ohms.

b) Calculate the power of the games console. *(2 marks)*

> *power = (current)² × resistance*
> *power = 5² × 10*
> *power = 25 × 10* ✓
> *power = 250 W* ✓

c) An update to the games console improves its efficiency. It now has a power of 230.4 W.
Assuming its resistance has not changed, what current is it now drawing? *(3 marks)*

> $current^2 = \dfrac{power}{resistance}$
> $current^2 = \dfrac{230.4}{10}$
> *current² = 23.04* ✓
> *current = √23.04* ✓
> *current = 4.8 A* ✓
>
> As the power rating has decreased whilst the resistance has remained the same, the current must be lower than 5 A.

d) A new games console draws a current of 15A and has a potential difference of 10V. Does this new console have a greater or smaller power than the updated console?
Write down any equation you use. *(4 marks)*

> power = potential difference × current ✓
> power = 10 × 15 ✓
> power = 150 W ✓
> This is smaller than the power of the updated console ✓.

This question requires you recall and apply a different power equation to the one in **a)**.

e) An update of the new console changes the current it draws so it now has a power of 125 W.
Assuming the potential difference is unchanged, what is the new current drawn by this games console? *(2 marks)*

> $current = \dfrac{power}{potential\ difference}$
> $current = \dfrac{125}{10}$ ✓
> current = 12.5 A ✓

Example

a) Which of the below does the amount of energy and appliance transfers **not** depend on? Tick **one** box. *(1 mark)*

How long the appliance is switched on for. ☐

The power of the appliance. ☐

The age of the appliance. ☐

> The age of the appliance. ✓

b) Whilst cooking a dish a microwave has a charge flow of 1130 coulombs and a potential difference of 120V.

i) Write down the equation that links energy transferred, potential difference and charge flow. *(1 mark)* energy transferred = charge flow × potential difference ✓

The units of this equation are energy transferred in joules, charge flow in coulombs and potential difference in volts.

ii) Calculate the energy transferred in the microwave.
Give your answer in kJ to three significant figures. *(2 marks)*

energy transferred = 1130 × 120
energy transferred = 135 600 J ✓
135 600 in kJ to 3 s.f. = 136 kJ ✓

iii) When cooking the same dish, a newer model of this microwave transfers 110 kJ while the potential difference remains unchanged.
What is the charge flow of the new model?
Give your answer to three significant figures. *(2 marks)*

$$\text{charge flow} = \frac{\text{energy transferred}}{\text{potential difference}}$$
$$\text{charge flow} = \frac{110\ 000}{120} ✓$$
$$\text{charge flow} = 917\ C ✓$$

As the energy transferred has decreased while the potential difference has remained the same, the charge flow must be below 1130 C. Convert the energy transferred into joules before substituting the value into the rearranged equation.

iv) The newer model of the microwave has a power of 1530 W. How long was the dish cooked for?
Give your answer in whole seconds and show how you arrived at your answer, including writing down any equations you use. *(3 marks)*

$$\text{time} = \frac{\text{energy transferred}}{\text{power}} ✓$$
$$\text{time} = \frac{110\ 000}{1530} ✓$$
$$\text{time} = 72\ s ✓$$

To answer this question you need to rearrange the energy transfer equation: energy transferred = power × time. Ensure you convert the energy transferred from kJ to J before substituting the value into the equation.

c) The table below shows the power ratings of different electrical appliances.

Appliance	Power rating
coffee maker	1400
dishwasher	1500
food blender	400
fridge / freezer	400
washing machine	500

Which of the appliances brings about the greatest transfer of energy in 10 minutes of use?
Explain your answer. *(2 marks)*

The dishwasher ✓, as it has the greatest power rating so would transfer the greatest amount of energy in the given time ✓.

Example

a) Which of the following statements is true? Tick **one** box. *(1 mark)*

Electrical power is transferred from power stations to consumers using the National Grid. ☐

Electrical power is transferred from consumers to power stations using the National Grid. ☐

Electrical force is transferred from power stations to consumers using the National Grid. ☐

Electrical power is transferred from power stations to consumers using the National Grid. ✓

b) Describe the use of step-up and step-down **transformers** in the National Grid. *(2 marks)*

Step-up transformers increase the potential difference from the power station to the transmission cables ✓. Step-down transformers are used to decrease the potential difference for use in homes ✓.

c) Explain why transformers in the National Grid are important for:

i) efficient transfer of energy *(2 marks)*

Step-up transformers increase potential difference for transmission by power cables as they reduce the current ✓, so reducing heat loss from the cables ✓.

ii) safe use of electricity. *(1 mark)*

Step-down transformers decrease the potential difference to a much lower level that is safe to use in appliances in the home ✓.

For more on the topics covered in this chapter, see pages 188–197 of the *Collins AQA GCSE Combined Science Revision Guide*.

Changes of State and the Particle Model

This short topic covers density and changes of state. You need to be able to recall and apply the pressure equation, explain the differences in density between the different **states of matter** in terms of the arrangement of **atom**s or molecules, and appreciate the differences between physical changes and chemical changes.

Example

a) Match the terms below with the correct **particle** model diagrams. *(2 marks)*

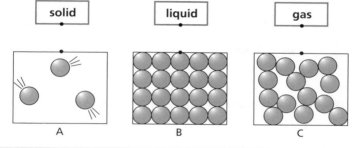

| solid | liquid | gas |

A   B   C

A: gas
B: solid
C: liquid
✓ ✓ One incorrect answer –1 mark.

b) Which of the substances pictured will have the lowest **density**? Explain how you arrived at your answer. *(2 marks)*

The gas (A) ✓, as the particles are the furthest apart ✓.

Example

a) Write down the equation that links density, mass and volume. *(1 mark)*

$$density = \frac{mass}{volume} \checkmark$$

The units of this equation are density in kg/m^3, mass in kg and volume in m^3.

b) Calculate the density of a rock sample with a mass of 6.7 kg and a
volume of 0.001 m³. *(2 marks)*

> $density = \frac{6.7}{0.001}$ ✓
>
> $density = 6700\,kg/m^3$ ✓

c) Another, more porous rock sample with the same volume has a density
of 5800 kg/m³.
What is the mass of this rock? *(2 marks)*

> $mass = density \times volume$
>
> $mass = 5800 \times 0.001$ ✓
>
> $mass = 5.8\,kg$ ✓

As this rock has a smaller density in the same volume than the rock in **b)** then the mass
value must be lower than 6.7 kg.

Example

a) Which of the following statements is correct? Tick **one** box. *(1 mark)*

Changes of state are a type of chemical change. ☐

Changes of state are physical changes. ☐

A material cannot recover its original properties after a physical change occurs. ☐

> *Changes of state are physical changes.* ✓

b) A 7 kg sample of water freezes completely to form ice.
What is the mass of the ice sample?
Explain how you arrived at your answer. *(2 marks)*

> *The ice has a mass of 7 kg* ✓; *this is because in a change of state the mass is conserved* ✓.

Water freezing is an example of a change of state that is a physical change as opposed
to a chemical change.

Internal Energy and Energy Transfers

The most important elements of this topic relate to changes in **internal energy**, particularly during changes of state. Make sure you don't mix up specific heat capacity and specific latent heat!

Specific heat capacity is the amount of energy required to raise the temperature of 1 kilogram of a substance by 1 degree Celsius. Specific latent heat is the amount of energy required to change the state of 1 kilogram of a substance with no change in temperature.

Example

An investigation was carried out into a solid substance that sublimes when heated. The solid was heated consistently for 10 minutes.

a) What **two** types of energy make up the internal energy of this system?   *(2 marks)*

> Kinetic energy ✓ and potential energy ✓.

> Internal energy is the energy stored inside a system by the particles that make up the system.

b) What happens to the internal energy of the system during this investigation? Explain your answer.   *(1 mark)*

> As it is being heated the internal energy increases ✓.

The graph below shows the results of the investigation.

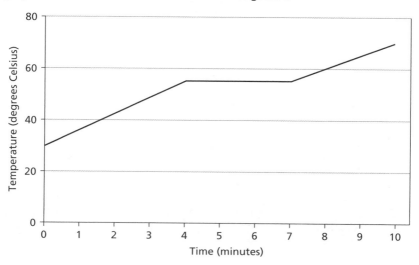

c) Explain the shape of the graph between 4 and 7 minutes. *(2 marks)*

> The temperature is constant at this point even though the substance is being heated ✓ because the substance is changing state from a solid to a gas ✓.

> When a substance sublimes it changes state from a solid to a gas.

Example

a) What is the difference between the specific **latent heat of fusion** and the specific **latent heat of vaporisation**? *(2 marks)*

> The specific latent heat of fusion is the energy required to change 1 kg of a substance from a solid to a liquid ✓. The specific latent heat of vaporisation is the energy required to change 1 kg of a substance from a liquid to a vapour ✓.

> The specific latent heat of a substance is the amount of energy required to change the state of 1 kg of the substance, with no change in temperature.

b) The specific latent heat of helium is 0.021 J/kg.
What energy would be required to change 780 g from a liquid to a vapour?
Use the correct equation from the Physics Equations Sheet.
Give your answer to two significant figures. *(2 marks)*

> energy for a change of state = mass × specific latent heat
>
> energy for a change of state = 0.78 × 0.021 ✓
>
> energy for a change of state = 0.016 J ✓

c) 0.35 J of energy is required to change the same mass of hydrogen from liquid to a vapour.
What is the specific latent heat of hydrogen?
Give your answer to two significant figures. *(2 marks)*

> $\text{specific latent heat} = \dfrac{\text{energy for a change of state}}{\text{mass}}$
>
> $\text{specific latent heat} = \dfrac{0.35}{0.78}$ ✓
>
> specific latent heat = 0.45 J/kg ✓

> As the energy value is greater than the answer for **b)**, this shows that the specific latent heat of hydrogen must be higher than that of helium.

Particle Model and Pressure

Example

An investigation into the properties of a gas was carried out. A sample of gas was sealed into a gas syringe.

a) The gas was heated, causing a rise in temperature from 45°C to 55°C. What effect would this have on the average kinetic energy of the molecules of the gas? *(1 mark)*

> The average kinetic energy of the molecules will increase ✓.

b) The gas remained at a volume of 130 cm³ while it was heated. What would happen to the **pressure** of the gas while it was heated? *(1 mark)*

> The pressure would increase ✓.

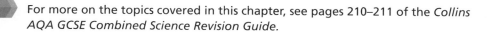

For more on the topics covered in this chapter, see pages 210–211 of the *Collins AQA GCSE Combined Science Revision Guide*.

Atoms and Isotopes

A key part of this topic is how the model of the atom has developed over time. You should be able to explain how new experimental evidence may lead to a scientific model being changed or replaced.

Example

a) Which of the below is the correct, approximate radius of an atom?
Tick **one** box. *(1 mark)*

1×10^{-2} metres ☐

1×10^{-20} metres ☐

1×10^{-10} metres ☐

1×10^{-10} metres ✓

The radius of a **nucleus** is less than $\frac{1}{10000}$ of the radius of an atom. Most of the mass of an atom is concentrated in the nucleus.

b) Use the answers in the box to complete the passage below. Each word can be used once, more than once or not at all. *(2 marks)*

positively   negatively   atom   nucleus   electrons   protons

The basic structure of an _____ is a _____

charged _____ composed of both _____ and

neutrons surrounded by _____ charged _____ .

The basic structure of an **atom** is a **positively** charged **nucleus** composed of both **protons** and neutrons surrounded by **negatively** charged **electrons**. ✓ ✓

One incorrect answer −1 mark.

c) The diagram below shows the structure of an atom.

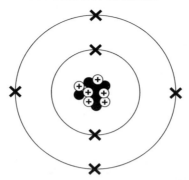

i) Explain how this structure would alter if the atom **absorbed** electromagnetic radiation. *(2 marks)*

> As electrons are absorbing electromagnetic radiation they would gain energy ✓ and jump to higher energy levels further from the nucleus ✓.

The electron arrangements of an atom may change with the absorption or the emission of electromagnetic radiation.

ii) All atoms of this **element** would have the same number of one of the particles found in the nucleus.
Identify this particle. *(1 mark)*

> All atoms of this element would have the same number of protons ✓.

The number of protons in an atom is called the **atomic number**.

iii) This element has an atomic number of 6 and a total of 12 particles in its nucleus.
How many neutrons are there in the nucleus? *(1 mark)*

> There are 6 neutrons in the nucleus ✓.

The number of protons + the number of neutrons = the mass number of the atom.

Atomic Structure

Example

Below are the three naturally occurring **isotopes** of hydrogen.

$$^{1}H \quad ^{2}H \quad ^{3}H$$

protium   deuterium   tritium

a) Explain the differences in the mass numbers. *(2 marks)*

> These isotopes have the same number of protons (one) and different
> numbers of neutrons ✓. Protium has zero neutrons, deuterium has one
> neutron, and tritium has two neutrons ✓.

> Atoms of the same element can have different numbers of neutrons, these atoms are
> called isotopes of that element.

b) Why are all of these atoms known as hydrogen when they have different
mass numbers? *(1 mark)*

> They all have the same number of protons in the nucleus, this makes them
> the same element ✓.

c) None of these isotopes are **ions**. Use this information to determine the number of
electrons in each isotope.
Explain how you arrived at your answer. *(2 marks)*

> Each isotope has one electron ✓, as they are not ions they are uncharged so
> must have the same number of protons as electrons ✓.

Example

Before the discovery of the electron, atoms were thought to be tiny spheres that
could not be divided.

a) Explain how the discovery of the electron led to this model being discarded and
the plum pudding model being adopted. *(3 marks)*

> The discovery of the electron meant that scientists realised that the
> atom contained smaller particles ✓. The plum pudding model suggested
> that the atom is a ball of positive charge ✓ with negative electrons
> embedded in it ✓.

The diagram below shows the results of the **alpha** particle scattering experiment carried out at Manchester University in 1909.

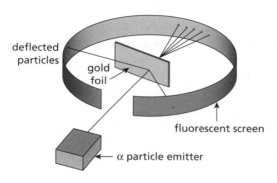

Rutherford, Geiger and Marsden's alpha scattering experiment

deflected particles

gold foil

fluorescent screen

α particle emitter

b) Using the diagram, explain how this experiment led to the rejection of the plum pudding model and the acceptance of a new model. *(4 marks)*

The alpha particle scattering experiment showed that most alpha particles were passing straight through the gold foil missing the nuclei of the atoms ✓, some were slightly deflected by passing close to the nuclei of the atoms ✓ and a small number were largely deflected due to striking the nucleus ✓. This led to the conclusion that the mass of an atom was concentrated in the centre at the nucleus and that the nucleus was charged. This is the nuclear model ✓.

There has been lots of further experimental work on the atom, including by Niels Bohr and James Chadwick.

c) Has the nuclear model been rejected due to this experimental work? Explain your answer, including examples. *(3 marks)*

No, the nuclear model has been modified ✓, e.g. Bohr's work led to the discovery that electrons orbit the nucleus at specific distances ✓. Chadwick's work provided the evidence to show the existence of neutrons within the nucleus ✓.

You are not required to know the details of Chadwick's experimental work or the experimental work supporting the Bohr model.

Atoms and Nuclear Radiation

This is a fairly wide-ranging topic covering a variety of different aspects of nuclear radiation. You should be able to balance nuclear equations, determine the **half-life** of a radioactive isotope from provided data, and explain the difference between radioactive contamination and irradiation.

Example

a) Explain the difference between **activity** and count-rate. *(2 marks)*

> Activity is the rate at which a source of **unstable** nuclei decays ✓.
> Count-rate is the number of decays recorded each second by a detector ✓.

> A Geiger-Muller tube is an example of a detector that can be used to measure count-rate.

b) An investigation was carried out into emission by **radioactive** sources. The radioactive sources **emitted** either alpha particles, **beta** particles or **gamma** rays. A detector was used to detect radiation emitted by different sources at a range of distances in the air. The results of the investigation are shown in the table below. A tick represents radiation being detected at a particular distance.

Source	Distance / cm			
	2	4	6	8
A	✓	✓	✗	✗
B	✓	✓	✓	✓
C	✓	✓	✓	✓
D	✓	✓	✗	✗

i) What conclusions can be drawn from the results of this investigation? Explain your answer. *(3 marks)*

> A and D are both emitting alpha particles ✓ as they are absorbed by a few centimetres of air so are not detected after 6 cm ✓. B and C could be emitting beta particles or gamma rays ✓.

ii) Design a follow-up investigation to determine the radiation emitted
by all of the sources. *(4 marks)*

> Set up a thin sheet of aluminium in front of sources B and C ✓. Place
> the detector on the other side of the aluminium sheet ✓. If radiation is
> detected on the other side of the aluminium sheet then the source must
> be emitting gamma rays ✓ as beta particles are absorbed by a thin
> sheet of aluminium ✓.

This follow-up investigation will require a method of distinguishing between beta
particles and gamma rays.

iii) What type of nuclear radiation emission is not being investigated
in this experiment? *(1 mark)*

> Neutron emission ✓

iv) What are the main hazards associated with this investigation? *(1 mark)*

> Irradiation or contamination from the radioactive sources ✓.

v) Outline one method of reducing these hazards. *(1 mark)*

> Ensure suitable protective clothing and equipment is used when
> handling the sources ✓.

Example

a) Identify the types of radioactive decay that would emit the particles
shown below. *(2 marks)*

A $\ _{-1}^{0}\text{e}$ B $\ _{2}^{4}\text{He}$

> A: beta decay ✓
> B: alpha decay ✓

An alpha particle is a helium nucleus and a beta
particle is an electron.

b) Explain the differing effects of alpha and beta decay on the mass and charge of the nucleus. *(4 marks)*

> Alpha decay causes both the mass and charge of the nucleus to decrease ✓ because two protons and two neutrons are released ✓. Beta decay does not change the mass of the nucleus as a neutron is converted to a proton ✓ but it does cause the charge of the nucleus to increase ✓.

c) What type of emission does not cause the mass or the charge of the nucleus to change? *(1 mark)*

> Gamma rays ✓

d) Complete the equations below.
Identify each example as either alpha or beta decay. *(3 marks)*

$$A \quad {}^{}_{38}\text{Sr} \longrightarrow {}^{90}_{}\text{Y} + {}^{0}_{-1}\text{e}$$

$$B \quad {}^{190}_{}\text{Pt} \longrightarrow {}^{}_{76}\text{Os} + {}^{}_{2}\text{He}$$

$$A \quad {}^{90}_{38}\text{Sr} \longrightarrow {}^{90}_{39}\text{Y} + {}^{0}_{-1}\text{e} \qquad \text{This is beta decay.}$$

$$B \quad {}^{190}_{78}\text{Pt} \longrightarrow {}^{186}_{76}\text{Os} + {}^{4}_{2}\text{He} \qquad \text{This is alpha decay.}$$

✓ ✓ ✓ −1 mark for each incorrect answer.

> You should be able to use the names and symbols of common nuclei and particles to write balanced equations that show single alpha (α) and beta (β) decay. You only need to be able to balance the atomic numbers and mass numbers, you won't need to identify the daughter elements (the elements that are produced by the decay).

Example

a) Use the answers in the box below to complete the following passage. Each answer can be used once, more than once, or not at all. *(2 marks)*

radiation	isotope	half-life	half	decay	gamma	nuclei

Radioactive _____ is random. The _____ of a

radioactive isotope is the time it takes for the number of _____

of the isotope in a sample to halve, or the time it takes for the count-rate

(or activity) from a sample containing the _____ to fall to

_____ its initial level.

> Radioactive **decay** is random. The **half-life** of a radioactive isotope is the time it takes for the number of **nuclei** of the isotope in a sample to halve, or the time it takes for the count-rate (or activity) from a sample containing the **isotope** to fall to **half** its initial level. ✓ ✓
>
> One incorrect answer –1 mark.

b) The graph below shows the decay of a radioactive isotope.

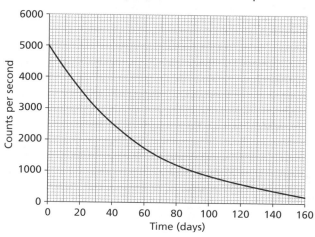

i) Estimate the half-life of this isotope.
Explain how you arrived at your answer. *(2 marks)*

> The half-life of this isotope is 40 days ✓, as this is the time it took for the activity (counts per second) to halve ✓.

ii) Estimate the counts per second of the sample after 160 days. *(1 mark)*

> counts per second after 80 days = $\frac{2500}{2}$ = 1250
>
> counts per second after 120 days = $\frac{1250}{2}$ = 625
>
> counts per second after 160 days = $\frac{625}{2}$ = 312.5 counts per second ✓

c) During the Chernobyl nuclear accident the town of Pripyat was covered in radioactive material ejected from the exposed core of the reactor of the nearby nuclear power plant. No one is able to live in Pripyat but people can visit and work there for short periods.

i) The Chernobyl disaster occurred in 1986. Explain why Pripyat is still dangerous. *(2 marks)*

> The area has become **contaminated** by the presence of radioactive atoms ✓, which are still decaying and emitting radiation ✓.

ii) Explain why people can spend short amounts of time in Pripyat but cannot live there permanently. *(2 marks)*

> People spending short periods of time at Pripyat are only receiving small doses of radiation ✓. People living there permanently would receive dangerously high doses of radiation over time ✓.

iii) A scientist visited Pripyat during a research trip. She was wearing a radiation detector on her coat. This showed her coat was absorbing nuclear radiation. When she returned to her lab she passed a radiation detector over the coat and the boots she was wearing. Her coat was not emitting radiation but her boots were. Explain these observations. *(2 marks)*

> Her coat had been **irradiated** whilst in Pripyat so was exposed to nuclear radiation but has not become radioactive ✓. Her boots must have become contaminated with radioactive material ✓.

iv) Why is it important that the scientist publishes any findings from her studies? *(1 mark)*

> So her results can be shared with other scientists and checked by peer review ✓.

For more on the topics covered in this chapter, see pages 212–217 of the *Collins AQA GCSE Combined Science Revision Guide*.

Forces and their Interactions

This topic covers the differences between **contact forces** and **non-contact forces**, **gravity** and resultant forces.

Example

Complete the table with answers from the box below. *(2 marks)*

| friction | tension | magnetic | gravitational | air resistance | electrostatic |

Contact forces	Non-contact forces

Contact forces	Non-contact forces
friction	gravitational
tension	magnetic
air resistance	electrostatic
✓	✓

A **contact force** is a force between objects that have to be physically touching. Non-contact forces can occur between objects that are physically separated.

Example

a) Write down the equation that links weight, mass and gravitational field strength.
(1 mark)

weight = mass × gravitational field strength ✓

The units of this equation are weight in newtons (N), mass in kg and gravitational field strength in N/kg.

b) Calculate the weight of a person with a mass of 74 kg. Assume gravitational field strength = 9.8 N/kg. *(2 marks)*

weight = 74 × 9.8 ✓
weight = 725.2 N ✓

In any calculation involving gravitational field strength the value will be given. The gravitational field strength on Earth is approximately 9.8 N/kg.

c) The gravitational field strength on Mars is approximately 3.8 N/kg.
How would the weight of the person in question **b)** differ if they were on Mars? Explain your answer. *(2 marks)*

Their weight would be lower ✓. As the gravitational field strength on Mars is lower than on Earth ✓.

Different planets have different gravitational field strength, which means that objects with the same mass would have different weights on different planets.

d) Calculate the mass of an object with a weight of 675 N on Mars. *(2 marks)*

$$mass = \frac{weight}{gravitational\ field\ strength}$$
$$mass = \frac{675}{3.8} ✓$$
$$mass = 177.6\ kg ✓$$

Rearrange the weight equation to find mass and use the gravitational field strength value from part **c)**.

e) What is the single point where the weight of an object acts called? *(1 mark)*

Centre of mass ✓

Example

The diagram below shows the forces acting on a moving truck.

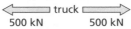

a) Calculate the **resultant** force on this truck. *(2 marks)*

500 − 100 ✓ = 400 kN ✓	In this question the resultant force is found by taking the larger force away from the smaller force. Make sure to use the correct units, which in this case are kN.

b) At a point later in the same journey the forces on the truck changed to those below.

⟸ truck ⟹
500 kN   500 kN

i) What statement can be made about the forces acting on the truck now? Explain your answer. *(2 marks)*

The forces on the truck are balanced ✓, as the forces are equal, so the resultant force on the truck is zero ✓.

Work Done and Energy Transfer

This short topic is mostly based around the work done equation. This is an equation you must be able to recall and apply.

Example

a) Write down the equation which relates work done, force and distance. *(1 mark)*

work done = force × distance

The units of this equation are work done in joules, force in newtons and distance in metres.

b) i) What work is done when a machine uses a force of 459 N to move object A 7 m? *(2 marks)*

work done = 459 × 7 ✓

work done = 3213 J ✓

ii) The settings of the machine are changed to move object B. The work done to move object B 4.2 m is 4918 J.
What is the force exerted on object B by the machine?   *(2 marks)*

$$force = \frac{work\ done}{distance}$$
$$force = \frac{4918}{4.2} \checkmark$$
$$force = 1171\,N \checkmark$$

Rearrange the equation from part **a)** so that force is the subject of the equation.

iii) Whilst object B is moving its temperature increases.
Explain why.   *(2 marks)*

Frictional forces act on the object as it moves ✓. *Work done against these forces causes a rise in temperature* ✓.

Unless an object is in a vacuum, friction will always be produced when an object moves.

c) What is 1318 Nm in kJ?
Give your answer to two significant figures.   *(1 mark)*

1318 Nm = 1.3 kJ ✓

1 joule (J) = 1 newton-metre (Nm). An answer to two significant figures should contain two digits that are not leading or trailing zeroes (zeroes at the start of the number).

Forces and Elasticity

This topic covers the effects of stretching, bending or compressing leading to elastic deformation or inelastic deformation.

Example

a) Write down the equation that links force applied to a spring, **spring constant** and extension.   *(1 mark)*

force applied to a spring = spring constant × extension ✓

The units of this equation are force in newtons (N), spring constant in newtons per metre (N/m) and extension in metres (m).

b) What force is needed to extend a spring 0.36 m where the spring has a spring constant of 65 N/m? *(2 marks)*

> $force = 65 \times 0.36$ ✓
>
> $force = 23.4\,N$ ✓

c) The same force was used to compress a large rubber ball which has a spring constant of 110 N/m.
What is the **compression** of the rubber ball? *(2 marks)*

> $\text{extension (compression)} = \dfrac{force}{spring\ constant}$
>
> $compression = \dfrac{23.4}{110}$ ✓
>
> $compression = 0.21\,m$ ✓

The equation from part **a)** can be used to calculate compression, where extension is the compression value. Rearrange the equation from **a)** to find the extension.

Example

The graphs below show the force over extension for two objects.

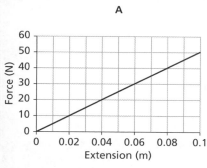

A

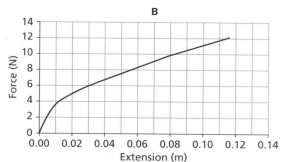

B

a) Which of the graphs show a linear relationship? *(1 mark)*

> Graph A ✓

Extension of an elastic object is directly proportional to the force applied.

b) Calculate the spring constant of the object in graph A. *(3 marks)*

> Values from graph:
> force = 10 N
> extension = 0.02 m ✓
> spring constant = $\frac{force}{extension}$
> spring constant = $\frac{10}{0.02}$ ✓
> spring constant = 500 N/m ✓

To calculate the spring constant take a force and extension value from the graph and rearrange the force extension equation to find the spring constant.

c) Would graph A continue with the same relationship if the force applied caused the limit of proportionality to be exceeded?
Explain your answer. *(2 marks)*

> No, it would no longer be a linear relationship ✓, as force and extension will no longer be proportional ✓.

d) What is the elastic potential energy of the object at 0.06 m on Graph A?
Show how you arrived at your answer. *(2 marks)*

> elastic potential energy = 0.5 × spring constant × (extension)²
> elastic potential energy = 0.5 × 500 × 0.06² ✓
> elastic potential energy = 0.9 N ✓

This question requires you to recall and apply the equation for elastic potential energy from the Energy topic.

e) What is the work done to extend the object?
Explain your answer. *(2 marks)*

> Work done = 0.9 N ✓. This is because the elastic potential energy stored is equal to work done if the spring is not **inelastically deformed** ✓.

Forces and Motion

This is a wide-ranging topic covering speed, velocity, acceleration and Newton's Laws of motion. Forces and braking covers the factors that affect the ability of a vehicle to come to a stop.

Example

a) Match the different activities to the typical speeds. *(2 marks)*

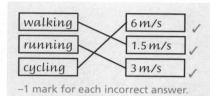

−1 mark for each incorrect answer.

b) A sound wave has a speed of 330 m/s in air.

i) Write down the equation that links distance, speed and time. *(1 mark)*

distance travelled = speed × time ✓

The units in this equation are distance in metres (m), speed in m/s and time in seconds (s).

ii) How far would the sound wave travel in 15 seconds? *(2 marks)*

distance travelled = 330 × 15 ✓
distance = 4950 m ✓

iii) The same sound wave would travel 22 km in water in the same time. What is the speed of sound in water? *(2 marks)*

speed = $\frac{distance}{time}$
speed = $\frac{22000}{15}$ ✓
speed = 1467 m/s ✓

Rearrange the equation so that speed is the subject of the equation. Make sure to convert 22 km into metres before substituting into the equation.

Forces

Example

A car accelerates from 15 m/s to 27 m/s in 4 seconds.

a) Write down the equation that links **acceleration**, change in **velocity** and time taken. *(1 mark)*

$$acceleration = \frac{change\ in\ velocity}{time\ taken} ✓$$

The units of this equation are acceleration in m/s², change in velocity in m/s and time taken in s.

b) Calculate the acceleration of the car. *(2 marks)*

$$acceleration = \frac{(27-15)}{4}$$
$$acceleration = \frac{12}{4} ✓$$
$$acceleration = 3\ m/s² ✓$$

To find the change in velocity take the final velocity away from the initial velocity.

c) The car changes speed from 29 m/s to 15 m/s with a **deceleration** of 2 m/s². What time did it take for this deceleration to occur? *(2 marks)*

$$time\ taken = \frac{change\ in\ velocity}{acceleration}$$
$$time\ taken = \frac{(15-29)}{-2} ✓$$
$$time\ taken = \frac{-14}{-2}$$
$$time\ taken = 7\ seconds ✓$$

As the car is going from a higher velocity to a lower velocity this is a deceleration, so will have a negative number for change in velocity.

Example

The graph below shows the motion of a boat.

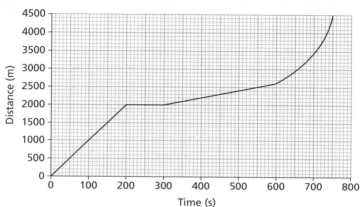

a) At what points is the boat stationary?
Explain your answer. *(2 marks)*

From 200 s to 300 s ✓, as the line is flat so the distance travelled by the boat is constant at this point ✓.

b) In which of the following time periods is the boat travelling the fastest?
Explain how you arrived at your answer. *(2 marks)*

0–200 s

300–600 s

0–200 s ✓, as this is the point where the line is the steepest ✓.

The speed of an object can be determined by the **gradient** of its distance–time graph.

c) What is the speed of the boat from 0 to 100 s?
Show your working. *(3 marks)*

change in distance = 1000 − 0 = 1000 m travelled

change in time = 100 − 0 = 100 s ✓

speed = $\frac{distance}{time}$

speed = $\frac{1000}{100}$ ✓

speed = 10 m/s ✓

Forces

Calculate the gradient of the line by dividing the change in the variable on the y-axis (in this case distance) by the corresponding change in the variable on the x-axis (in this case time).

d) HT How could you determine the speed at 700 s? *(2 marks)*

Draw a tangent to the curve at 700 s ✓ and find the gradient of the tangent. This will be the speed of the boat ✓.

The line is curving upwards at 700 s so the boat is accelerating.

Example

a) Explain why speed is a **scalar** quantity whilst velocity is a **vector** quantity. *(2 marks)*

Speed only has magnitude so is a scalar quantity ✓. Velocity has a magnitude and a direction so is a vector quantity ✓.

b) A motorbike accelerates from 0 to 30 m/s in 6 s.
Use the correct equations from the Physics Equations Sheet to calculate the distance it covers in this time. *(5 marks)*

(final velocity)^2 – (initial velocity)^2 = 2 × acceleration × distance

First, calculate the acceleration of the motorbike:

$$acceleration = \frac{change\ in\ velocity}{time\ taken}$$

$$acceleration = \frac{(30-0)}{6}$$

$$acceleration = \frac{30}{6}\ ✓$$

$$acceleration = 5\ m/s^2\ ✓$$

Now find the result of the left side of the equation:

(final velocity)^2 – (initial velocity)^2 = $30^2 - 0^2$

(final velocity)^2 – (initial velocity)^2 = 900 ✓

900 = 2 × acceleration × distance

Substitute the acceleration value (5 m/s²) and rearrange the equation to make distance the subject.

distance = 900 ÷ 2 ÷ 5 ✓

distance = 90 m ✓

Example

The graph below shows the velocity of a runner during a race.

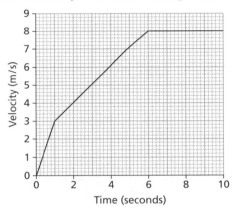

Time (seconds)

a) What was the runner's acceleration between 2 and 6 seconds? *(2 marks)*

change in velocity = 8 − 4 = 4 s ✓

change in time = 6 − 2 = 4 s

acceleration = $\frac{4}{4}$ = 1 m/s² ✓

The acceleration of an object can be calculated from the gradient of a velocity–time graph.

b) HT What distance does the runner travel between 6 and 10 s? *(2 marks)*

area under graph from 6 to 10 s = (10 − 6) × (8 − 0)

area under graph from 6 to 10 s = 4 × 8 ✓

area under graph from 6 to 10 s = 32 m ✓

The distance travelled by an object (or **displacement** of an object) can be calculated from the area under a velocity–time graph.

c) i) `HT` What distance does the runner run between 2 and 6 s? *(2 marks)*

> *area under graph from 2 to 6 s = 0.5 × (6−2) × (8−4) + (6−2) × (4−0)* ✓
> *area under graph from 2 to 6 s = 8 + 16*
> *area under graph from 2 to 6 s = 24m* ✓

ii) `HT` Explain the difference in your answers to **b)** and **c) i)**. *(2 marks)*

> *As the runner's velocity was greater at 6−10 s than at 2−6 s* ✓,
> *the distance covered in the same length of time (4 s) was greater* ✓.

Forces, Accelerations and Newton's Laws of Motion

Example

Use the answers in the box to match up Newton's Laws of motion with the correct description. *(2 marks)*

First Law	Second Law	Third Law

Description	Law
If the resultant force acting on an object is zero and the object is stationary, the object remains stationary.	
Whenever two objects interact, the forces they exert on each other are equal and opposite.	
The acceleration of an object is proportional to the resultant force acting on the object.	

Description	Law
If the resultant force acting on an object is zero and the object is stationary, the object remains stationary.	First Law
Whenever two objects interact, the forces they exert on each other are equal and opposite.	Third Law
The acceleration of an object is proportional to the resultant force acting on the object.	Second Law

✓ ✓ One incorrect answer −1 mark.

Example

A cyclist is riding down a mountain road.

a) The cyclist is maintaining a constant velocity.
What statement can be made about the restive force and the driving force
of the cyclist? *(1 mark)*

> They are balanced ✓.

b) The cyclist begins pedalling faster and accelerates.
What statement can be made about the restive force and driving force now?
Use the term 'resultant force' in your answer. *(2 marks)*

> The driving force is greater than the resistive force ✓, this means that there
> is now a resultant force on the cyclist ✓.

c) **HT** The cyclist comes to a stop and will remain in this state of rest until she
accelerates again. What term is given to the tendency of objects to continue
in their state of rest? *(1 mark)*

> Inertia ✓

Example

A skier (A) who has a mass of 90 kg accelerates at 3 m/s².

a) Write down the equation that links resultant force, mass and acceleration. *(1 mark)*

> resultant force = mass × acceleration ✓

The units of this equation are force in N, mass in kg and acceleration in m/s².

b) What is the resultant force on the skier? *(2 marks)*

> resultant force = 90 × 3 ✓
> resultant force = 270 N ✓

c) A second skier (B) accelerates at the same rate but has a resultant force acting on
them of 240 N. What is their mass? *(2 marks)*

> mass = $\frac{resultant\ force}{acceleration}$
> mass = $\frac{240}{3}$ ✓
> mass = 80 kg ✓

As the resultant force is lower, this means the
mass of skier B must be lower than that of the
first skier.

Forces

HT d) If the skiers are travelling at the same speed, skier A will have the greater inertial mass.

What is meant by the term inertial mass? *(1 mark)*

> Inertial mass is a measure of how difficult it is to change the velocity of an object ✓.

Example

a) What **two** distances make up the stopping distance? *(1 mark)*

> The thinking distance and the **braking distance** ✓.

The thinking distance is the distance the vehicle travels during the driver's reaction time. The braking distance is the distance the vehicle travels under the braking force.

b) Which of the below values is a typical human reaction time? Tick **one** box. *(1 mark)*

0.02 seconds ☐

0.2 seconds ☐

2 seconds ☐

> 0.2 seconds ✓

c) Which of the following factors would **not** affect the thinking distance? Tick **two** boxes. *(1 mark)*

The condition of the tyres ☐

The driver having consumed alcohol ☐

The driver being tired ☐

A wet road ☐

> The condition of the tyres; A wet road ✓

d) On French motorways the speed limit is 130 km/h in the dry and 110 km/h when it is raining. Use the idea of stopping distance to explain why this is the case. *(3 marks)*

> On a wet road stopping distance increases due to the braking distance increasing ✓. Speed also affects stopping distance so a car travelling at 130 km/h would have a much longer stopping distance on a wet road than a car travelling at 110 km/h ✓. As the stopping distance is shorter, travelling at 110 km/h reduces the chance of accidents occurring ✓.

You should relate your answer to the specific context of the question, including using the **two** speed values in your answer.

Example

a) Explain why a vehicle stops when the brakes are applied. *(3 marks)*

> When a force is applied to the brakes, work done ✓ by the friction force between the brakes and the wheel ✓ reduces the kinetic energy of the vehicle until it reaches zero and the vehicle stops ✓.

b) Many high-performance cars have specially designed air-flow systems to maximise the flow of air over the brakes. This is particularly important during large decelerations.
Explain why. *(4 marks)*

> During large decelerations, work done by the friction force of the brakes against the wheels ✓ leads to large increases in the temperature of the brakes ✓. This can lead to the brakes overheating and stop working ✓. By maximising the air flow over the brakes they will be cooled down during large decelerations ✓.

Example

Reaction time can be measured using an app on a smartphone. When the colour of the screen changes the subject must press a button as quickly as possible. The time it takes them to press the button is then displayed by the app.

Design an investigation into the effect of distraction on thinking distance. Suggest methods to ensure the results are not seriously affected by random error and **anomalous** results are correctly dealt with. *(6 marks)*

First have the subject complete the test multiple times; this will allow them to get used to the test and reduce the effect of learning on the results. Once their reaction times are consistent, begin recording the times. Make sure they do the test in a silent room with no distractions. After a set number of tests, e.g. 20, give the subject a short break. The subject should now repeat the test the same number of times but during each test they should be distracted in the same way by being talked to. Record their reaction time after each test. Once you have completed your investigation discard any results that are very different from the others (anomalous results). Calculate a mean reaction time for completing the test undistracted and a mean reaction time for completing the test whilst being distracted.

This is an example of a 6-mark extended response question. It is marked in levels. The table below shows the different level descriptors. In order for your answer to be placed in a level you have to satisfy the criteria laid out for that particular level. The quality of your answer then determines what mark you are awarded within that level. The key aspect of all these questions is developing a sustained line of reasoning which is coherent, relevant, substantiated and logically structured. This means you should focus on writing well structured answers with a logical order that relate directly to the question and do not contain any non-relevant material.

When designing an investigation in an exam ensure your instructions are clear and logically ordered, your method ensures the experiment is **reproducible** and you've taken steps to reduce random and systematic errors.

Level	Marks
Level 3: Clear and coherent description of investigation method including correct description of discarding anomalous results and calculation of a mean.	5–6
Level 2: Partial description of investigation, which may not be logically ordered. Correct reference made to anomalous results but lacking in a clear explanation of how to deal with them.	3–4
Level 1: Basic description of simple method with incorrect or very underdeveloped reference to anomalous results.	1–2
No relevant content.	0

Momentum

Students often find **momentum** a challenging topic but the key thing to remember is that in a closed system total momentum before an event is equal to total momentum after an event. This is conservation of momentum.

Example

A car with a mass of 1300 kg is travelling at 15 m/s during a crash test.

a) Write down the equation that relates momentum, mass and velocity. *(1 mark)*

> *momentum = mass × velocity* ✓
>
> The units for this equation are momentum in kg m/s, mass in kg and velocity in m/s.

b) Calculate the momentum of the car. *(2 marks)*

> *momentum = 1300 × 15* ✓
> *momentum = 19 500 kg m/s* ✓

c) The car changed velocity and its momentum changed to 24 700 kg m/s. What was the car's new velocity? *(2 marks)*

> $velocity = \frac{momentum}{mass}$
> $velocity = \frac{24\,700}{1300}$ ✓
> $velocity = 19\,m/s$ ✓
>
> Rearrange the equation from part a) so that velocity is the subject.

d) The car hits another car, which is stationary. What is the total momentum of the two cars when they collide?

Explain your answer. *(3 marks)*

> *Momentum is conserved in a* **collision** ✓*. The momentum of the moving car is 24 700 kg m/s. As the other car is stationary, its momentum is 0* ✓*. Total momentum = 24 700 + 0 = 24 700 kg m/s* ✓*.*

For more on the topics covered in this chapter, see pages 158–169 of the *Collins AQA GCSE Combined Science Revision Guide*.

Waves in Air, Fluids and Solids

This topic looks at a number of different characteristics of waves, including the relationship between speed, frequency and **amplitude**.

Example

a) Use the answers in the box below to complete the table on the features of a wave.   *(2 marks)*

wavelength	amplitude	frequency

Definition	Key feature
The number of waves passing a point each second.	
The maximum displacement of a point on a wave away from its undisturbed position.	
The distance from a point on one wave to the equivalent point on the adjacent wave.	

Definition	Key feature
The number of waves passing a point each second.	frequency
The maximum displacement of a point on a wave away from its undisturbed position.	amplitude
The distance from a point on one wave to the equivalent point on the adjacent wave.	wavelength

✓ ✓ One incorrect answer −1 mark.

b) Which of the following statements is not a feature of **longitudinal waves**?
Tick **one** box. *(1 mark)*

Show areas of compression and rarefaction. ☐

The **oscillations** of the wave are perpendicular to the direction of energy transfer. ☐

The oscillations of the wave are parallel to the direction of energy transfer. ☐

The oscillations of the wave are perpendicular to the direction of energy transfer. ✓

Example

A group of school students were carrying out an investigation into waves. They were using a ripple tank and a stroboscope.

a) What type of waves were they investigating? *(1 mark)*

The students were investigating water waves ✓.

b) The students placed a ping-pong ball in the tank. The ball floated and moved up and down as the wave passed, but did not move along the tank. Explain the property of waves that this provides evidence of. *(2 marks)*

As the ball only moved up and down when the water wave passed, this shows that the waves transfer energy and information ✓ *but do not transfer matter* ✓.

Example

a) The frequency of a wave is 250 Hz. What is the **period** of this wave? Write down any equations you use. *(2 marks)*

$period = \frac{1}{frequency}$ ✓

$period = \frac{1}{250}$

$period = 0.004 \text{ s}$ ✓

b) Write down the equation that links wave speed, frequency and wavelength. *(1 mark)*

$wave\ speed = frequency \times wavelength$ ✓

The units of this equation are wave speed in m/s, frequency in Hz and wavelength in m.

c) What is the speed of a wave that had a frequency of 3 kHz and a wavelength of 0.03 m? *(2 marks)*

> wave speed = 3000 × 0.03 ✓
>
> wave speed = 90 m/s ✓

Convert 3 kHz into Hz before substituting the value into the equation. There are 1000 Hz in 1 kHz.

Electromagnetic Waves

Electromagnetic waves are **transverse** waves that transfer energy from the source of the waves to an absorber. Electromagnetic waves form a continuous spectrum and all types of electromagnetic wave travel at the same velocity through a vacuum or air.

Example

a) Complete the electromagnetic spectrum using the answers in the box below.
(2 marks)

microwaves	ultraviolet	radio waves

long wavelength ──────────────→ short wavelength

A	B	infrared	visible light	C	X-rays	gamma rays

A = radio waves

B = microwaves

C = ultraviolet ✓ ✓

One incorrect answer −1 mark.

b) Which type of electromagnetic radiation is detected by our eyes? *(1 mark)*

> Visible light ✓

Example

The diagram below shows the refraction of a light ray as it travels between two media.

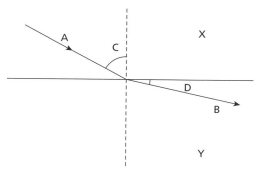

a)   Identify A to D.   *(4 marks)*

> A = incident ray ✓
> B = **refracted** ray ✓
> C = angle of incidence ✓
> D = angle of refraction ✓

b)   **HT** Which of the media (X or Y) is the most optically **dense**?
>   Explain how you arrived at your answer.   *(2 marks)*

> Media X is the most dense ✓, as the light ray bent away from the **normal**
> as it moved from media X to media Y ✓.

> Light bends towards the normal when entering a more optically dense **medium** and
> bends away from the normal when entering a less optically dense medium.

c) The diagram below shows a wave moving between air and water.
Use the diagram to explain why the wave has refracted. *(3 marks)*

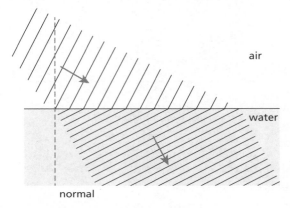

air

water

normal

> As the wave travels slower in a denser medium ✓, the edge of the wave that
> hits the water first slows down while the rest of the wave continues at the
> same speed ✓. This causes the wave to bend towards the normal ✓.

Example

In an experiment to investigate radio waves an antenna was set up to detect the
radio waves.

a) HT Explain how radio waves are produced. *(1 mark)*

> Radio waves can be produced by oscillations in electrical circuits ✓.

b) HT The antenna is connected to a circuit that can measure the frequency of an
alternating current.
Explain how this apparatus can be used to measure the frequency of
radio waves. *(2 marks)*

> When radio waves are absorbed they may create an alternating current ✓
> with the same frequency as the radio wave ✓.

c) Use the words in the box below to complete the following passage. Each word can be used once, more than once, or not at all. *(2 marks)*

| ultraviolet | gamma | waves | nucleus | atoms |

Changes in _____ and the nuclei of atoms can result in

electromagnetic _____ being generated or absorbed over a

wide frequency range. _____ rays originate from changes in the

_____ of an atom.

Changes in **atoms** and the nuclei of atoms can result in electromagnetic **waves** being generated or absorbed over a wide frequency range. **Gamma** rays originate from changes in the **nucleus** of an atom. ✓ ✓
One incorrect answer −1 mark.

For more on the topics covered in this chapter, see pages 182–187 of the *Collins AQA GCSE Combined Science Revision Guide*.

Permanent and Induced Magnetism, Magnetic Forces and Fields

This is a fairly straightforward topic area covering the key points of magnets and magnetic fields.

Example

An investigation was carried out into the properties of two **permanent magnets**. When one **pole** of one magnet was brought together with the pole of another magnet they **repelled** each other.

a) What statement could be made about these two poles? *(1 mark)*

> *These poles must be like, e.g. north and north or south and south ✓.*
>
> The two poles on a magnet are north and south.

b) This is an example of a non-contact force.
Explain why. *(1 mark)*

> *The magnets did not have to be touching in order to experience the force ✓.*
>
> Forces can be separated into contact forces and non-contact forces.

c) Strips of magnetic material were used in this investigation. Which of the following is not a magnetic material? Tick **one** box. *(1 mark)*

iron ☐
aluminium ☐
cobalt ☐
nickel ☐

> *aluminium ✓*

> Four magnetic materials are listed in the specification – you should make sure you learn all four of these.

d) A strip of magnetic material was brought into one of the bar magnet's magnetic fields. Predict the force this strip would experience.
Explain your answer. *(2 marks)*

> It always experiences a force of **attraction** ✓, as the strip is an **induced magnet** and induced magnetism always causes a force of attraction ✓.

The strip is an induced magnet.

e) Whilst in the magnetic field of the bar magnet the strip could be used to pick up small magnetic objects. Explain what would happen to the strip's magnetism if it was removed from the bar magnet's magnetic field. *(1 mark)*

> The strip was an induced magnet so when removed from a magnetic field it loses its magnetism quickly ✓.

f) The magnetic strip was placed at different points around the bar magnet. At which part of the magnet would the strip experience the strongest force? Explain your answer. *(2 marks)*

> The strip would feel the strongest force at the poles of the magnet ✓. This is the point where the magnetic field of the magnet is the strongest ✓.

The strongest force will be experienced in the area where the magnetic field of the magnet is the strongest.

g) As part of the investigation a compass was used to plot the magnetic field pattern around the bar magnet.
Explain how this could be carried out. *(3 marks)*

> Place the compass inside the magnetic field and the compass will point to the south pole of the magnet ✓. Draw a small line in the direction the compass is pointing ✓. Repeat for different points around the magnet ✓.

Magnetic field lines run from the north pole to the south pole of a magnet.

h) Explain why a compass can be used in navigation. *(2 marks)*

> A compass points in the direction of the Earth's magnetic field ✓. This means it will always point north ✓.

Use the fact that the Earth has a magnetic field in your answer.

i) What evidence does a compass provide about the core of the Earth? *(1 mark)*

> It provides evidence that the core of the Earth is magnetic and so generates a magnetic field ✓.

The Motor Effect

This is a more challenging topic, including applying knowledge of magnetic fields to practical applications of the **motor effect**.

Example

An investigation was carried out into the magnetic field produced by a **current** flowing through a conducting wire. A magnetic object was placed in the magnetic field.

a) Predict the effect of the following on the strength of the attraction experienced by the magnetic object.
In each case explain your answer.

i) Increasing the distance from the wire. *(2 marks)*

> Increasing the distance from the wire would decrease the force of attraction experienced ✓, as the strength of the magnetic field would decrease ✓.

ii) Increasing the amplitude of the current flowing through the wire. *(2 marks)*

> Increasing the amplitude of the current flowing through the wire would increase the force of attraction experienced ✓, as the strength of the magnetic field would increase ✓.

Distance from the wire and the amplitude of the current flowing through the wire are two of the factors that affect the strength of a magnetic object placed in the magnetic field of an **electromagnet**.

b) What effect would shaping the wire to form a **solenoid** have on the magnetic field? *(1 mark)*

It would increase the strength of the magnetic field ✓.

A solenoid is formed by wrapping the wire carrying the current into a coil.

c) A compass was used to plot the shape of the magnetic field around the solenoid. What statement could be made about the shape of the plot produced? *(1 mark)*

It would be the same shape as the magnetic field around a bar magnet ✓.

d) An iron core was added to the solenoid. What was now being produced? *(1 mark)*

An electromagnet ✓

e) The magnetic field was once again plotted. How would the plot have changed? Explain your answer. *(2 marks)*

The field lines would be closer together ✓, *as the magnetic field would be stronger* ✓.

Placing an iron core into a solenoid increases the strength of the magnetic field.

HT Example

During a school physics practical into the motor effect, a simple motor was constructed using a magnet and a wire.

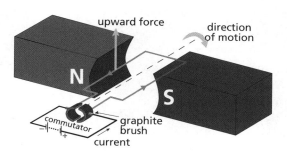

a) Explain what will happen when a current is passed through the wire. *(1 mark)*

The magnet and the conductor will exert a force on each other ✓.

Magnetism and Electromagnetism

b) Describe the effect of the following on the size of the force experienced by the conductor.

  i) Decreasing the strength of the magnetic field. *(1 mark)*

   Decreasing the strength of the magnetic field will decrease the force experienced by the conductor ✓.

  ii) Increasing the current flowing through the wire. *(1 mark)*

   Increasing the current flowing through the wire will increase the force experienced by the conductor ✓.

   The strength of the magnetic field and the current flowing through the wire are two factors which affect the size of the force.

c) What rule links the relative orientation of the force, the current in the conductor and the magnetic field? *(1 mark)*

   Fleming's left-hand rule ✓

In this investigation the magnetic **flux density** was 2.1 T, the current was 14 A and the conductor had a length of 10 cm.

d) What was the force experienced by the conductor? Use the correct equation from the Physics Equations Sheet. *(2 marks)*

   force = magnetic flux density × current × length
   force = 2.1 × 14 × 0.1 ✓ force = 2.94 N ✓

   Convert the length of the conductor into metres before substituting the value into the equation.

e) The current was changed and a force of 1.89 N was experienced by the conductor. What is the new current? *(2 marks)*

   current = force ÷ (magnetic flux density × length)
   current = 1.89 ÷ (2.1 × 0.1) ✓
   current = 9 A ✓

   As the force has decreased the current must also have decreased.

For more on the topics covered in this chapter, see pages 206–209 of the *Collins AQA GCSE Combined Science Revision Guide*.

Glossary

Biology

Abiotic factor non-living component of the ecosystem

Active immunity production of antibodies by introducing a pathogen into the body

Active site area on an enzyme (lock) that a substrate molecule (key) can fit into

Active transport the movement of substances against a concentration gradient; requires energy

Amino acids building block molecules that link together to form proteins

Amylase an enzyme that breaks down starch

Antibiotic medicine / drug produced to combat bacterial infections

Antibody specific protein produced in response to a specific antigen on a pathogen

Antigen protein on the outside of a foreign cell / particle that can be recognised by antibodies

Asexual reproduction produces new individuals that are identical to their parents; does not involve the fusion of gametes

Auxins a group of growth hormones produced in plants

Binomial system the method of naming organisms by using their genus and species

Biotic factor living component of the ecosystem

Carriers (genetics) heterozygous individuals who possess a recessive gene (causing a disorder) but do not exhibit symptoms as they also have a dominant gene that 'masks' them

Cilia microscopic hairs found on the surface of cells

Cystic fibrosis a genetic disorder resulting in excessive production of thick mucus and difficulty releasing digestive enzymes

Denatured where the 3-D structure of a protein is changed, usually by heat

Differentiation process by which cells become specialised. Most cells in the embryonic state go through this process

Dominant an allele that always expresses itself

Double circulatory system blood circulation divided into two – one pulmonary (to the lungs) and the other systemic (to the rest of the body)

Droplet infection a mode of disease transmission by aerosol, e.g. coughing and sneezing to produce droplets containing pathogens

Effectors organs (usually muscles) that respond to impulses from motor neurones – in the case of muscles, by contracting

Fertile the ability of an organism to reproduce successfully by sexual means

Fossil fuel coal, oil, natural gas or peat, formed from fossilisation of plant material and can be burned

Gamete a specialised sex cell formed by meiosis

Glossary

Genotype the combination of alleles an individual has for a particular gene, e.g. BB, Bb or bb

Haploid a chromosome set that is half in number, typically found in gametes produced by meiosis

Heterozygous when an individual carries two different alleles for a gene, e.g. Bb

Homozygous when an individual carries two copies of the same allele for a gene, e.g. BB or bb

Ligase enzyme an enzyme used in genetic engineering to 'splice' sections of DNA together

Limiting factors in photosynthesis, where a factor, e.g. carbon dioxide concentration, limits the rate of photosynthesis despite another factor being increased

Malignant of cancer, where cells spread from the original site of a tumour to other parts of the body

Meiosis cell division that forms daughter cells with half the number of chromosomes of the parent cell

Memory cells cells in the immune system that remain dormant in the body and are sensitised (can recognise) a particular pathogen if it invades

Mimicry anti-predator strategy where an organism copies the body structure or behaviour of another to deter the predator

Mitosis cell division that forms two daughter cells, each with the same number of chromosomes as the parent cell

Molecules a collection of atoms joined together by bonds

Mutation where the sequence of bases in DNA is changed and produces a new protein structure

Natural selection process by which organisms evolve (proposed by Charles Darwin)

Non-coding base sequences non-sense DNA – some play a role in the switching on of genes during protein production

Osmosis the movement of water, through a partially permeable membrane, into a solution with a lower water concentration

Pathogen a disease-causing microorganism

Pentadactyl limb five-membered limb, e.g. wing, hand, flipper

Phenotype the physical expression of the genotype, i.e. the characteristic shown

Photosynthesis a process in green plants by which sunlight energy is used to synthesise carbohydrate using carbon dioxide and water

Pituitary a small gland at the base of the brain that produces hormones; known as the 'master gland'

Plasmolysis state in plant cells where the cytoplasm or protoplast pulls away from the cell wall due to loss of water

Polydactyly a genetic condition caused by a dominant allele, where affected people have extra fingers or toes

Potometer apparatus used to measure rates of transpiration in plants

Protein large polymer molecule formed from chains of amino acids

Purified free from contaminants. One substance only

Recessive an allele that will only be expressed if there are two present; represented by a lower case letter

Respiration process in all living cells whereby energy is released from the breakdown of energy-rich molecules such as glucose

Restriction enzyme in genetic engineering, an enzyme that removes specific segments of DNA

Ribosome sub-cellular structure where proteins are made

Statins medicines that reduce cholesterol build-up in the body

Stent cylindrical structure placed in arteries during surgery to maintain a wider diameter and unrestricted blood flow

Stroke volume measurement of the volume of blood output from the heart

Transpiration process of water loss from a plant via the leaves

Turgor pressure in plant cells; the outward pressure on the cell wall resulting from cytoplasm and vacuole gaining or holding water

Variation differences between individuals of the same species

Zygote a fertilised ovum or egg cell. A zygote is also a stem cell

Chemistry

Activation energy the minimum amount of energy required for a reaction to take place

Algae small green aquatic plants

Alkanes saturated hydrocarbons with the general formula C_nH_{2n+2}

Alkenes unsaturated hydrocarbons with the general formula C_nH_{2n}

Alloy a mixture that contains at least one metal, e.g. steel

Anode a positively charged electrode

Aqueous dissolved in water

Atom the smallest part of an element that can exist

Atomic number the number of protons in the nucleus of an atom

HT Avogadro constant the number of particles in one mole of any substance, i.e. six hundred thousand billion billion or 6.02×10^{23}

Billion 1 000 000 000 or one thousand million

HT Bioleaching the use of bacteria to extract metals from low-grade ores

Carbon footprint the total amount of carbon dioxide and other greenhouse gases that are emitted over the full life cycle of a product, a service or an event

Catalyst a substance that speeds up the rate of a chemical reaction, but is not used up in the reaction itself

Cathode a negatively charged electrode

Cell (electrical) contains chemicals that react together to release electricity

Cellulose a form of carbohydrate

Chromatography a separation technique used to separate the coloured components of mixtures

Compound a substance containing atoms of two or more elements, which are chemically combined in fixed proportions

Concentration The amount of substance in a given volume, normally measured in units of g/dm^3

Conservation of mass the total mass of the products of a chemical reaction is always equal to the total mass of the reactants

Glossary

Covalent bond a shared pair of electrons between atoms in a molecule

Cracking the process by which longer-chain hydrocarbons can be broken down into shorter, more useful hydrocarbons

Crude oil a fossil fuel; a mixture consisting mainly of alkanes

Cryolite a compound of aluminium

Crystallisation technique used to obtain a soluble solid from a solution

Delocalised not bound to one atom

Desalinated salt water that has had all the dissolved salts removed to make pure water

Diamond a form of carbon, with a giant covalent structure, that is very hard

Displacement reaction a reaction in which a more reactive metal displaces a less reactive metal from a solution of its salt

Dissociate to split up into ions

Electrode an electrical conductor used in a cell

Electrolysis the process by which an ionic compound is broken down into its elements using an electrical current

Electron a subatomic particle with a relative mass that is very small and a relative charge of −1

Electron configuration represents how the electrons are arranged in shells around the nucleus of an atom

Electrostatic a force of attraction between oppositely charged species

Element a substance made of only one type of atom

Empirical formula the simplest whole number ratio of the atoms of each element present in a compound

Endothermic reaction a reaction that takes in energy from the surroundings

Equation a scientific statement that uses chemical names or symbols to sum up what happens in a chemical reaction

Equilibrium A reversible reaction where the rate of the forward reaction is equal to the rate of the reverse reaction

Exothermic reaction a reaction that gives out energy to the surroundings

Extraction the process of obtaining / taking out / removing, e.g. obtaining a metal from an ore or compound

Fermentation a chemical reaction in which alcohol and carbon dioxide are produced from glucose

Filtration separation techniques used to separate insoluble solids from soluble solids

Formulation a mixture that has been carefully designed to have specific properties

Fractional distillation a separation technique used to separate mixtures that contain components with similar boiling points

Fullerene a form of carbon in which the carbon atoms are joined together to form hollow structures, e.g. tubes, balls and cages

HT Gradient = difference in the y-axis value ÷ difference in the x-axis value

Graphene a form of carbon; a single layer of graphite, just one atom thick

Graphite a form of carbon, with a giant covalent structure, that conducts electricity

Greenhouse gases gases (e.g. carbon dioxide, water vapour or methane) that, when present in the atmosphere, absorb the outgoing infrared radiation, increasing the Earth's temperature

HT **Half equation** used to show what happens to one of the reactants in a chemical reaction

Halide a halogen ion, e.g. fluoride, chloride, bromide or iodide

Halogens elements in Group 7 of the Periodic Table

Hydrocarbon a molecule that contains only carbon and hydrogen atoms

Indicator a chemical that is one colour in an acid and another colour in an alkali

Infrared a type of radiation that is part of the electromagnetic spectrum

Intermolecular between molecules

Ion atoms that have gained or lost electrons and now have an overall charge

Ionic bond the force of attraction between positive and negative ions

HT **Ionic equation** a simplified version of a chemical equation that just shows the species that are involved in the reaction

HT **Ionised** split up into ions

Isotope atoms of the same element that have the same atomic number but a different mass number

HT **Le Chatelier's Principle** a rule that can predict the effect of changing conditions on a system that is in equilibrium

Life cycle assessment (LCA) used to assess the environmental impact a product has over its whole lifetime

HT **Limiting reactant** the reactant that is completely used up in a reaction; it stops the reaction from going any further and any further products being produced

Malleable can be hammered into shape

Mass number the total number of protons and neutrons in the nucleus of an atom

Matter describes the material that everything is made up of

Melting point the specific temperature at which a pure substance changes state, from solid to liquid and from liquid to solid

Mendeleev, Dmitri the Russian chemist who devised the Periodic Table

Metallic bond the strong attraction between metal ions and delocalised electrons

Mixture a combination of two or more elements or compounds, which are not chemically combined together

Mobile phase in chromatography, the phase that does move (e.g. the solvent)

HT **Mole (mol)** the unit for measuring the amount of substance; one mole of any substance contains the same number of particles; the mass of one mole of a substance is equal to the relative formula mass in grams

Molecule two or more atoms covalently bonded together

Molten liquefied by heat

Monomer a small molecule that can join together to form a polymer

Neutralisation occurs when an acid reacts with exactly the right amount of base

Glossary

Neutron a subatomic particle with a relative mass of 1 and no charge

Noble gases unreactive non-metal elements that have a full outer shell of electrons and are found in Group 0 of the Periodic Table

Oxidation when a species loses electrons

Particle a small piece of matter

Photosynthesis a chemical reaction carried out by green plants in which carbon dioxide and water react to produce glucose and oxygen

HT **Phytomining** the use of plants to extract metals from low-grade ores

Polymers large molecules formed when lots of smaller monomer molecules join together

Potable water that is safe to drink

Precipitate a solid formed when two solutions are combined

Products the substances produced in a chemical reaction

Proton a subatomic particle with a relative mass of 1 and relative charge of +1

Pure a substance that contains only one type of element or compound

Rate of reaction a measure of how quickly a reactant is used up or a product is made

Reactants the substances that react together in a chemical reaction

Reaction profile (energy level diagram) a diagram showing the relative energies or the reactants and the products of a reaction, the activation energy and the overall energy change of the reaction

Reactivity series a list in which elements are placed in order of how reactive they are

Reduction when a species gains electrons

Relative atomic mass (A_r) the ratio of the average mass per atom of the element to one-twelfth the mass of an atom of carbon-12

Relative formula mass (M_r) the sum of the relative atomic masses of all the atoms shown in the formula

Reversible reaction a reaction that can go forwards or backwards

Saturated a hydrocarbon molecule that only contains single carbon–carbon (C–C) bonds

Silicon dioxide (silica) (SiO_2) a compound found in sand

Simple distillation a technique used to separate a substance from a mixture due to a difference in the boiling points of the components in the mixture

HT **Species** the different atoms, molecules or ions that are involved in a reaction

Stationary phase in chromatography, the phase that does not move (e.g. the absorbent paper)

HT **Strong acid** an acid, such as hydrochloric acid, nitric acid or sulfuric acid, that is completely ionised in water

HT **Tangent** a straight line that touches a curve at one point

HT **Titration** an accurate technique that can be used to find out how much of an acid is needed to neutralise an alkali

Transition metals metals that have typical metal properties and are found in the middle of the Periodic Table

Unsaturated a hydrocarbon molecule with at least one double carbon–carbon (C=C) bond

Viscous sticky

HT **Weak acid** an acid, such as ethanoic acid, citric acid or carbonic acid, that is only partially ionised in water

Physics

Absorb / absorbed to take in and retain (all or some) incident radiated energy

Acceleration / accelerating the rate of change of velocity, measured in metres per second squared (m/s^2)

Activity the rate at which a radioactive source emits radiation, measured in becquerel

Alpha a type of radiation, which is strongly ionising, in the form of a particle consisting of two neutrons and two protons (a helium nucleus)

Alternating current (a.c.) a continuous electric current that periodically reverses direction, e.g. mains electricity in the UK

Amplitude the maximum displacement that any particle in a wave achieves from its undisturbed position, measured in metres (m)

Anomalous a data value that is significantly above or below the expected value / outside a pattern or trend

Atom the smallest quantity of an element that can take part in a chemical reaction, consisting of a positively charged nucleus made up of protons and neutrons, surrounded by negatively charged electrons

Atomic number the number of protons in an atom of an element

Attraction a force by which one object attracts ('pulls') another, e.g. gravitational or electrostatic force

Beta a type of nuclear radiation that is moderately ionising, consisting of a high-speed electron, which is ejected from a nucleus as a neutron turns into a proton

Braking distance the distance a vehicle travels under braking force (from the point when the brakes are first applied to the point when the vehicle comes to a complete stop)

Charge a property of matter that causes it to experience a force when placed in an electric field; electric current is the flow of charge; charge can be positive or negative and is measured in coulombs (C)

HT **Collision** an event in which two or more bodies, or particles, come together with a resulting change of direction and energy

Compression the act of squeezing / pressing (an elastic object)

Conductivity a measure of the ability of a substance to conduct electricity

Contact force a force that occurs between two objects that are in contact (touching), e.g. friction and tension

Contamination / contaminated the unwanted presence of materials containing radioactive atoms on other materials

Current the flow of electrical charge, measured in amperes (A)

Deceleration describes negative acceleration, i.e. when an object slows down

Dense having a high density (mass per volume)

Density a measure of mass per unit of volume, measured in kilograms per metre cubed (kg/m^3)

Glossary

Diode a component that only allows current to flow in one direction (has a very high resistance in the reverse direction)

Direct current (d.c.) a continuous electric current that flows in one direction only, without significant variation in magnitude, e.g. the current supplied by cells and batteries

Displacement a vector quantity that describes how far and in what direction an object has travelled from its origin in a straight line

Distance a scalar quantity that provides a measure of how far an object has moved (without taking into account direction)

Efficiency the ratio or percentage of useful energy out compared to total energy in for a system or device

Elastic potential energy the energy stored in a stretched / compressed elastic object, such as a spring

Electromagnet a magnet consisting of an iron or steel core wound with a coil of wire, through which a current is passed

Electromagnetic (EM) waves a continuous spectrum of waves formed by electric and magnetic fields, ranging from high frequency gamma rays to low frequency radio waves

Electron a subatomic particle, with a charge of −1, which orbits the nucleus of an atom

Element a substance that consists only of atoms with the same number of protons in their nuclei

Emit / emitted to give off (radiation or particles)

Energy a measure of the capacity of a body or system to do work

Extension the distance over which an object (such as a spring) has been extended / stretched

Fluid a substance, such as a liquid or a gas, which can flow; has no fixed shape

Flux density a measure of the density of the field lines around a magnet

Force an influence that occurs when two objects interact

Frequency the number of times that a wave / vibration repeats itself in a specified time period

Fusion a reaction in which two nuclei combine to form a nucleus with the release of energy

Gamma high frequency, short wavelength electromagnetic waves; a type of nuclear radiation, emitted from a nucleus

Gradient a measure of the steepness of a sloping line; the ratio of the change in vertical distance over the change in horizontal distance

Gravitational potential energy (GPE) the energy gained by raising an object above ground level (due to the force of gravity)

Gravity the force of attraction exerted by all masses on other masses, only noticeable with a large body, e.g. the Earth or Moon

Half-life the average time it takes for half the nuclei in a sample of radioactive isotope to decay; the time it takes for the count-rate / activity of a radioactive isotope to fall by 50% (halve)

Induced magnet an object that becomes magnetic when placed in a magnetic field

Inelastically deformed describes an object that cannot return to its original shape when the forces that caused it to change shape are removed (because the limit of proportionality has been exceeded)

HT **Inertia** the tendency of a body to stay at rest or in uniform motion unless acted upon by an external force

Infrared the part of the electromagnetic spectrum with a longer wavelength than light but a shorter wavelength than radio waves

Internal energy the sum of the energy of all the particles that make up a system, i.e. the total kinetic and potential energy of all the particles added together

Ion formed when an atom loses or gains one or more electrons to become charged

Irradiation / irradiated to expose an object to nuclear radiation (the object does not become radioactive)

Isotope atoms of the same element that have different numbers of neutrons

Kinetic energy the energy of motion of an object, equal to the work it would do if brought to rest

Latent heat of fusion the amount of heat energy needed for a specific amount of substance to change from solid to liquid

Latent heat of vaporisation the amount of heat energy needed for a specific amount of substance to change from liquid to gas

Limit of proportionality the point up to which the extension of an elastic object is directly proportional to the applied force (once exceeded the relationship is no longer linear)

Longitudinal wave a wave in which the oscillations are parallel to the direction of energy transfer, e.g. sound waves

Mass a measure of how much matter an object contains, measured in kilograms (kg)

Medium a material or substance

Microwaves electromagnetic radiation in the wavelength range 0.3 to 0.001 metres, used in satellite communication and cooking

HT **Momentum** the product of an object's mass and velocity

HT **Motor effect** the force experienced by a current carrying conductor when it is placed in a magnetic field, which is used to create movement in an electrical motor

National Grid the network of high voltage power lines and transformers that connects major power stations, businesses and homes

Neutron a neutral subatomic particle; a type of nuclear radiation, which can be emitted during radioactive decay

Non-contact force a force that occurs between two objects that are not in contact (not touching), e.g. gravitational and electrostatic forces

Normal at right angles to / perpendicular to

Nucleus the positively charged, dense region at the centre of an atom, made up of protons and neutrons, orbited by electrons

Ohmic conductor a resistor in which the current is directly proportional to the potential difference at a constant temperature

Oscillate / oscillations to vibrate / swing from side to side with a regular frequency

Parallel (circuit) a circuit in which the components are connected side by side on a separate branch / path, so that the current from the cell / battery splits with a portion going through each component

Glossary

Particle an extremely small body with finite mass and negligible (insignificant) size, e.g. protons, neutrons and electrons

Period the time taken for a wave to complete one oscillation; the time it takes for a particle in the wave to move backwards and forwards once around its undisturbed position

Permanent magnet an object that produces its own magnetic field

Pole the two opposite regions in a magnet, where the magnetic field is concentrated; can be north or south

Potential difference the difference in electric potential between two points in an electric field; the work that has to be done in transferring a unit of positive charge from one point to another, measured in volts (V)

Power a measure of the rate at which energy is transferred or work is done

Pressure the force exerted on a surface, e.g. by a gas on the walls of a container

Proportional describes two variables that are related by a constant ratio

Proton a subatomic particle found in the nucleus of an atom, with an electrical charge of +1

Radioactive containing a substance that gives out radiation

Refracted when a wave meets a boundary between two different materials and changes direction

Renewable can be replaced

Reproducible results are reproducible if the investigation / experiment can be repeated by another person, or by using different equipment / techniques, and the same results are obtained, demonstrating that the results are reliable

Repulsion / repelled a force that pushes two objects apart, such as the force between two like electric charges or magnetic poles

Resistance a measure of how a component resists (opposes) the flow of electrical charge, measured in ohms (Ω)

Resultant a single force that represents the overall effect of all the forces acting on an object

Scalar a quantity, such as time or temperature, that has magnitude but no direction

Series (circuit) a circuit in which the components are connected one after the other, so the same current flows through each component

Solenoid formed by coiling a wire to increase the strength of the magnetic field created by a current through the wire

Specific heat capacity the amount of energy required to raise the temperature of one kilogram of substance by one degree Celsius

Speed a scalar measure of the distance travelled by an object in a unit of time, measured in metres per second (m/s)

Spring constant a measure of how easy it is to stretch or compress a spring; calculated as: force ÷ extension

State of matter the structure and form of a substance, i.e. gas, liquid or solid

System an object or group of objects

Transferred refers to how energy is changed, e.g. chemical energy can be transferred to electric energy

Transformer a device that transfers an alternating current from one circuit to another, with an increase (step-up transformer) or decrease (step-down transformer) of voltage

Transverse a wave in which the oscillations are at right angles to the direction of energy transfer, e.g. water waves

Ultraviolet the part of the electromagnetic spectrum with wavelengths shorter than light but longer than X-rays

Universe all existing matter, energy and space

Unstable lacking stability; having a very short lifetime; radioactive

Vector a variable quantity that has magnitude and direction

Velocity a vector quantity that provides a measure for the speed of an object in a given direction

Wavelength the distance from one point on a wave to the equivalent point on the next wave, measured in metres (m), represented by the symbol λ

Weight the vertical downwards force acting on an object due to gravity

Work the product of force and distance moved along the line of action of the force, when a force causes an object to move

X-rays the part of the electromagnetic spectrum with wavelengths shorter than that of ultraviolet radiation but longer than gamma rays

Chemistry Formula Sheet

You are not provided with a formula sheet in your chemistry exams. Some questions may provide formulae but it is safest to learn all of the ones that are mentioned in the specification.

You will also be expected to rearrange formulae to change the subject of the equation.

Equation	Notes
number of neutrons = mass number – atomic number	If you are asked to calculate the number of neutrons in an atom, then you will be provided with the mass number and the atomic number for the atom. It may be in the following form: $$^{23}_{11}Na$$ In this format, the top number is the mass number. You may find it helpful to remember that the 'massive' number is the 'mass' number.
law of conservation of mass: $$\text{mass of reactants} = \text{mass of products}$$	In a balanced chemical equation, the total mass of products formed is equal to the total mass of the reactants used.
HT $$\text{moles} = \frac{\text{mass}}{M_r}$$	Use this equation when you are calculating the number of moles from a mass. M_r is the relative formula mass of the compound. It is the sum of the relative atomic masses of the atoms shown in the formula. You will also be expected to know how to rearrange this equation to make mass and M_r the subject of the equation.

$\text{concentration} = \dfrac{\text{mass}}{\text{volume}}$	The units of concentration in this case will be g/dm^3. The volume refers to the liquid that the solid has been dissolved in and needs to be in dm^3.
HT energy transferred = bonds broken – bonds formed	This equation is used when bond energies are supplied. A negative answer means that the reaction is exothermic and a positive answer means the reaction is endothermic.
$\text{mean rate of reaction} = \dfrac{\text{quantity of reactant used}}{\text{time taken}}$	This equation and the one in the row below can be used to calculate the mean rate of a reaction.
HT $\text{mean rate of reaction} = \dfrac{\text{quantity of product formed}}{\text{time taken}}$	The units of rate of reaction may be given as g/s or cm^3/s. The units may also be given in mol/s.
HT $\text{gradient} = \dfrac{\text{vertical length}}{\text{horizontal length}}$	To calculate the rate of reaction at time t: • draw a tangent to the curve at time t • draw a right-angled triangle with the tangent as the hypotenuse • use the scale on each axis to find the triangle's vertical length and horizontal length. The gradient of the curve will tell you the rate of reaction at time t.
$R_f = \dfrac{\text{distance moved by substance}}{\text{distance moved by solvent}}$	Use this equation when you are asked to calculate the R_f value for a compound in chromatography.

Physics Formulae and Physics Equations Sheet

When solving quantitative problems, you should be able to recall and apply the following equations using standard SI units. This means these equations will **not** be given to you on the exam papers.

Word equation	Symbol equation
weight = mass × gravitational field strength (g)	$W = m\,g$
work done = force × distance (along the line of action of the force)	$W = F\,s$
force applied to a spring = spring constant × extension	$F = k\,e$
distance travelled = speed × time	$s = v\,t$
acceleration = $\dfrac{\text{change in velocity}}{\text{time taken}}$	$a = \dfrac{\Delta v}{t}$
resultant force = mass × acceleration	$F = m\,a$
HT momentum = mass × velocity	$p = m\,v$
kinetic energy = 0.5 × mass × (speed)²	$E_k = \dfrac{1}{2}\,m\,v^2$
$\begin{array}{l}\text{gravitational}\\ \text{potential energy}\end{array}$ = mass × $\begin{array}{l}\text{gravitational}\\ \text{field strength}\end{array}$ (g) × height	$E_p = m\,g\,h$
power = $\dfrac{\text{energy transferred}}{\text{time}}$	$P = \dfrac{E}{t}$
power = $\dfrac{\text{work done}}{\text{time}}$	$P = \dfrac{W}{t}$
efficiency = $\dfrac{\text{useful output energy transfer}}{\text{total input energy transfer}}$	
efficiency = $\dfrac{\text{useful power output}}{\text{total power input}}$	
wave speed = frequency × wavelength	$v = f\,\lambda$
charge flow = current × time	$Q = I\,t$
potential difference = current × resistance	$V = I\,R$
power = potential difference × current	$P = V\,I$
power = (current)² × resistance	$P = I^2\,R$

Physics Formulae and Physics Equations Sheet

energy transferred = power × time	$E = P\,t$
energy transferred = charge flow × potential difference	$E = Q\,V$
density = $\dfrac{\text{mass}}{\text{volume}}$	$\rho = \dfrac{m}{V}$

You should be able to select and apply the following equations from the Physics Equations Sheet, which will be given to you as part of the exam papers.

Word equation	Symbol equation
(final velocity)² − (initial velocity)² = 2 × acceleration × distance	$v^2 - u^2 = 2\,a\,s$
elastic potential energy = 0.5 × spring constant × (extension)²	$E_e = \frac{1}{2}\,k\,e^2$
change in thermal energy = mass × specific heat capacity × temperature change	$\Delta E = m\,c\,\Delta\theta$
period = $\dfrac{1}{\text{frequency}}$	
HT force on a conductor (at right angles to a magnetic field) carrying a current = magnetic flux density × current × length	$F = B\,I\,l$
thermal energy for a change of state = mass × specific latent heat	$E = m\,L$
HT potential difference across primary coil × current in primary coil = potential difference across secondary coil × current in secondary coil	$V_p\,I_p = V_s\,I_s$

Periodic Table

Key

Metals

Non-metals

Relative atomic mass →		
Atomic symbol →	**H**	
Name →	hydrogen	
Atomic / proton number →	1	

Group 1	2											3	4	5	6	7	0 or 8
																	4 **He** helium 2
7 **Li** lithium 3	9 **Be** beryllium 4											11 **B** boron 5	12 **C** carbon 6	14 **N** nitrogen 7	16 **O** oxygen 8	19 **F** fluorine 9	20 **Ne** neon 10
23 **Na** sodium 11	24 **Mg** magnesium 12											27 **Al** aluminium 13	28 **Si** silicon 14	31 **P** phosphorus 15	32 **S** sulfur 16	35.5 **Cl** chlorine 17	40 **Ar** argon 18
39 **K** potassium 19	40 **Ca** calcium 20	45 **Sc** scandium 21	48 **Ti** titanium 22	51 **V** vanadium 23	52 **Cr** chromium 24	55 **Mn** manganese 25	56 **Fe** iron 26	59 **Co** cobalt 27	59 **Ni** nickel 28	63.5 **Cu** copper 29	65 **Zn** zinc 30	70 **Ga** gallium 31	73 **Ge** germanium 32	75 **As** arsenic 33	79 **Se** selenium 34	80 **Br** bromine 35	84 **Kr** krypton 36
85 **Rb** rubidium 37	88 **Sr** strontium 38	89 **Y** yttrium 39	91 **Zr** zirconium 40	93 **Nb** niobium 41	96 **Mo** molybdenum 42	[98] **Tc** technetium 43	101 **Ru** ruthenium 44	103 **Rh** rhodium 45	106 **Pd** palladium 46	108 **Ag** silver 47	112 **Cd** cadmium 48	115 **In** indium 49	119 **Sn** tin 50	122 **Sb** antimony 51	128 **Te** tellurium 52	127 **I** iodine 53	131 **Xe** xenon 54
133 **Cs** caesium 55	137 **Ba** barium 56	139 **La*** lanthanum 57	178 **Hf** hafnium 72	181 **Ta** tantalum 73	184 **W** tungsten 74	186 **Re** rhenium 75	190 **Os** osmium 76	192 **Ir** iridium 77	195 **Pt** platinum 78	197 **Au** gold 79	201 **Hg** mercury 80	204 **Tl** thallium 81	207 **Pb** lead 82	209 **Bi** bismuth 83	[209] **Po** polonium 84	[210] **At** astatine 85	[222] **Rn** radon 86
[223] **Fr** francium 87	[226] **Ra** radium 88	[227] **Ac*** actinium 89	[261] **Rf** rutherfordium 104	[262] **Db** dubnium 105	[266] **Sg** seaborgium 106	[264] **Bh** bohrium 107	[277] **Hs** hassium 108	[268] **Mt** meitnerium 109	[271] **Ds** darmstadtium 110	[272] **Rg** roentgenium 111	[285] **Cn** copernicium 112	[286] **Uut** ununtrium 113	[289] **Fl** flerovium 114	[289] **Uup** ununpentium 115	[293] **LV** livermorium 116	[294] **Uus** ununseptium 117	[294] **Uuo** ununoctium 118

*The lanthanides (atomic numbers 58–71) and the actinides (atomic numbers 90–103) have been omitted.

The relative atomic masses of copper and chlorine have not been rounded to the nearest whole number.